Following the Gun:

Enforcing Federal Laws Against Firearms Traffickers

June 2000

Department of the Treasury
Bureau of Alcohol, Tobacco & Firearms

FOREWORD BY THE DIRECTOR

Virtually every crime gun in the United States starts off as a legal firearm. Unlike narcotics or other contraband, the criminals' supply of guns does not begin in clandestine factories or with illegal smuggling. Crime guns, at least initially, start out in the legal market, identified by a serial number and required documentation. This means that virtually every crime gun leaves some paper trail. Historically, the Bureau of Alcohol, Tobacco and Firearms (ATF) has pursued cases against both armed offenders and firearms traffickers. Until recently, however, we did not have the tools to develop a systematic approach to understanding and addressing the sources of firearms used in crime. In 1996, we intensified our efforts to address the illegal supply of crime guns by using the crime gun recovered by law enforcement officials to identify and target the illegal suppliers. To assess this strategy, we have conducted a review of our trafficking investigations and their disposition by prosecutors and courts.

Studies such as this one are essential to ATF. The last decade has brought our investigators and inspectors an enormous increase in investigative information. The National Tracing Center (NTC) now has over 1,000,000 traces of firearms recovered by law enforcement officials in our firearms trafficking information system. The National Integrated Ballistics Information Network (NIBIN), operated by ATF with the Federal Bureau of Investigation (FBI), now contains 500,000 ballistics images. The National Instant Check System (NICS), launched by the FBI and ATF in 1998, has resulted in ATF receiving over 130,000 reports of prohibited persons attempting to buy firearms from FFLs. Our files contain facts from thousands of investigations and debriefings of arrested persons in possession of firearms used in crime.

This huge advance in investigative information and in tools to access it enables ATF and our State and local partners to identify many more criminals individually, and to analyze and respond to specific local crime patterns. With greater knowledge of the gun criminal, we can target our resources more effectively and better explain our cases to the communities we serve. At the same time, this wealth of criminal investigative information brings with it the management challenges of creating the best investigative and strategic uses, sharing information with other Federal, State, and local authorities, protecting citizen privacy, and fully informing Congress and the public.

The case analysis presented here will help us develop the most effective possible enforcement strategies. This report demonstrates the effectiveness of State and local law enforcement agencies and prosecutors joining ATF in "following the crime gun" to the gun's illegal supplier, and targeting that supplier and others in the chain of illegal transfers. It may be the gun of the drug dealer, the violent gang member, the repeat felon, parolee or probationer, the domestic violence offender, the juvenile, or any other person prohibited from possessing a firearm. Gun traffickers are often criminals in other respects, and trafficking investigations provide another means to prevent them from harming the community.

This report is strong evidence of the support ATF receives from the U.S. Attorneys and other prosecuting attorneys on a daily basis. Our review would not have been possible without the assistance from valuable academic partners and the Bureau of Justice Statistics, as well as support from the Department of the Treasury. We also want to thank the many State and local law enforcement agencies who directly participated in the vast majority of investigations described in this report. Every member of the study team joins me in expressing our gratitude.

Bradley A. Buckles

Table of Contents

LIST OF TABLES

APPENDICES

EXECUTIVE SUMMARY

Many criminals obtain their guns from the illegal market supplied by a variety of sources: unlicensed sellers who buy guns with the purpose of reselling them; fences; corrupt Federal firearms licensees (FFLs); and straw purchasers who buy guns for other unlicensed sellers, criminal users, and juveniles. It is the responsibility of the Bureau of Alcohol, Tobacco and Firearms (ATF), often working together with State and local law enforcement agencies, to investigate this criminal trafficking in firearms, arrest the perpetrators, and refer them to U.S. Attorneys for prosecution.

To report on the problem and the Federal enforcement response to it, ATF documented and analyzed the criminal investigations involving firearms traffickers that it initiated from July 1996 through December 1998, from commencement of the investigation to sentencing by a court. This review builds on an earlier examination of the illegal acquisition of firearms by youth and juveniles, prepared in response to a request by Congress.[1] The information for both reports derived from surveys of the ATF special agents responsible for the investigations.[2]

The analysis documents an aggressive, productive effort that led to the prosecution, conviction, and sentencing of hundreds of firearms traffickers during this period. It also suggests that this effort could be rendered still more effective with continued improvements in investigative techniques and enforcement tools.

The investigations and the trafficked firearms

Trafficking and armed offender investigations. ATF initiated 1,530 firearms trafficking investigations during the period July 1996 through December 1998, which comprised only a portion of ATF's firearms investigations. Most ATF firearms investigations were initiated specifically to target offenders who committed acts of armed violence or were considered potentially violent — armed career criminals, armed narcotics traffickers, and felons in possession of firearms — as part of locally designed strategic efforts to reduce drug trafficking and violent gang activity, and violent crime generally. ATF trafficking investigations complement these enforcement strategies by reducing the illegal availability of firearms to such armed violent offenders as well as to juveniles, and by identifying and arresting other violent criminals through investigations of firearms trafficking activity.

Removing guns from the streets. The targets of the ATF trafficking investigations during this period diverted a total of 84,128 firearms from legal to illegal commerce. ATF estimates that agents seized a quarter of these firearms in connection with the investigations themselves, removing them from the illegal market. The remainder of the trafficked firearms were documented during the criminal investigation or later recovered in crimes. Trafficked firearms may continue to be recovered by law enforcement officials and traced long after traffickers are arrested and incarcerated.

How ATF initiates trafficking investigations. Most of the trafficking investigations were initiated based on traditional case methods, such as referrals from other agencies and information provided by confidential informants. One in every five investigations, however, was triggered by information provided by Federal firearms licensees through tips or mandatory reporting to ATF concerning lost or stolen firearms.

---

[1] *Performance Report* for the Senate and House Committee on Appropriations, Youth Crime Gun Interdiction Initiative, Department of the Treasury, Bureau of Alcohol, Tobacco and Firearms, February 1999.

[2] The survey was designed by Dr. Anthony A. Braga, of the John F. Kennedy School of Government, Harvard University, who also prepared the tables in this report, in consultation with ATF. For a full description of the methodology, see Appendix B.

Role of firearms tracing. Almost 30 percent of the investigations (448 of 1,530) were initiated through the innovative investigative methods of information analysis – analysis of firearms trace data, multiple firearms sales records, or both. After initiation of the investigations, tracing was used as an investigative tool to gain information on recovered crime guns in 60 percent of the investigations (918 of 1,530). These findings show that crime gun trace analysis and on-line Project LEAD, the National Tracing Center's firearms trafficking information system, are being widely used to increase ATF's productivity in the development of firearms trafficking investigations. Almost 60 percent of the investigations involved secondhand guns, which are very difficult to trace because unlicensed sellers are not required to keep any transfer records, and there is no effective way to track a gun beyond the first retail sale. Investigative use of firearms and ballistics trace information by Federal, State, and local law enforcement to identify traffickers and armed offenders is a developing capability that should be strengthened.

Traffickers and armed criminals

Traffickers supply criminals and juveniles with guns. Traffickers move guns onto the streets and into the hands of criminals. Over half of the trafficking investigations involved firearms known to have been subsequently involved in additional criminal investigations, including investigations of homicide and robbery, assault, felon in possession of firearms, and illegal gun possession. Juveniles were involved in about 14 percent of the investigations (209) as possessors, thieves and robbers, and traffickers. The number of crime guns linked to particular traffickers by ATF in investigations or through firearms and ballistics tracing is likely to understate significantly the role of trafficked firearms in violent crime. Investigative resources for trafficking cases are limited, and firearms and ballistics tracing are not yet fully developed and available to law enforcement agencies.

Trafficking investigations lead to armed violent felons. When law enforcement follows the gun used by the criminal to its illegal supplier, the investigation often leads to another violent criminal. Convicted felons play a significant role in firearms trafficking. About 23 percent of the investigations included violations involving convicted felons buying, selling, or possessing firearms. A quarter (669) of the traffickers identified in the investigations were convicted felons. About 45 percent of the trafficking investigations involved convicted felons in various roles, including trafficking and receiving trafficked firearms.

Traffickers and trafficking channels

Different types of traffickers and trafficking channels. Firearms traffickers are using a variety of channels to divert firearms, and investigations usually involve multiple trafficking channels, such as a corrupt FFL and a straw purchaser, or theft and unlicensed dealing.

Corrupt FFLs as major traffickers. Although FFLs were involved in under 10 percent of the trafficking investigations, they were associated with the largest number of diverted firearms – over 40,000 guns, nearly half of the total number of trafficked firearms documented during the two-year period. The mean number of trafficked guns involved in any case in which an FFL figured was over 350. When an FFL was acting as the sole trafficker in the investigation, or working with an unlicensed dealer, the mean number of guns per investigation rose to over 550. Clearly, FFLs' access to large numbers of firearms makes them a particular threat to public safety when they fail to comply with the law. Investigations focused on retail gun stores, pawnshops, and residential FFLs. The 133 investigations of FFLs revealed a variety of violations, including failure to keep required records, transfers to prohibited persons, offenses involving National Firearms Act weapons, making false entries in record books, and conducting illegal out-of-state transfers.

Gun shows. Gun shows were a major trafficking channel, involving the second highest number of trafficked guns per investigation (more than 130), and associated with approximately 26,000 illegally diverted firearms. The investigations involved both licensed and unlicensed sellers at gun shows.

Straw purchasers. Straw purchasing was the most common channel in trafficking investigations. Almost half of all the trafficking investigations involved straw purchasers. Therefore, although the average number of firearms trafficked per straw purchase investigation was relatively small, 37 firearms, there were nearly 26,000 firearms associated with these investigations.

Unlicensed sellers. Unlicensed sellers were a focus of about a fifth of the trafficking investigations, and involved an average of about 75 guns per investigation and almost 23,000 guns. Unlicensed sellers range from individuals who knowingly sell guns to criminals from their personal collections to interstate gun runners buying guns to sell to gangs and drug organizations.

Firearms theft. Firearms theft is an important source of trafficked firearms. Firearms stolen from FFLs, residences, and common carriers were involved in over a quarter of the trafficking investigations. Investigations involving firearms stolen from residences and federally licensed firearms dealers were associated with over 9,000 trafficked firearms. There were a handful of investigations involving thefts of firearms from common carriers, but such thefts may involve a large number of firearms. ATF is proposing a regulation that would require FFLs to report guns missing in shipment.

Productive partnerships between ATF, U.S. Attorneys, and State and local authorities

Successful prosecutions. Over 1,000 of ATF's trafficking investigations were referred to prosecutors. Prosecutors accepted 90 percent of ATF's trafficking case referrals, involving more than 1,700 defendants. At the time of the survey, over 60 percent of the defendants were fully adjudicated. Over 1,000 traffickers — 97 percent of those adjudicated — were found guilty and sentenced in Federal, State, and local courts. Over three-quarters received sentences of incarceration up to life in prison; most of the remainder received probation. Altogether, as a result of these investigations, 812 defendants were sentenced in Federal court to a cumulative 7,420 years in prison, with an average sentence of about nine years.

State and local law enforcement role. More than 33,000 offenders were convicted of felony firearms-related offenses in State and local courts in 1996.[3] State or local law enforcement agencies participated in 68 percent (1,037 of 1,530) of the ATF trafficking investigations. Although only about 3 percent of ATF's case referrals and about 10 percent of the defendants referred were to State and local prosecutors, about 70 percent of ATF's investigations involved intrastate trafficking and about 10 percent involved guns stolen from residences. This suggests that where there are applicable laws, State and local law enforcement play an important role in curbing the illegal market in firearms.

Enforcement challenges. Investigators and prosecutors face challenges building successful cases against firearms traffickers. Prosecutors accepted over 60 percent of trafficking cases referred by ATF that involved straw purchasers and dealing by unlicensed sellers, but in the majority of cases proceeded with other charges, not related on their face to firearms trafficking. Also notable is that although FFLs are associated with the largest volume of trafficked firearms, many FFL violations are misdemeanors rather than felonies, presenting a dilemma for prosecutors who

---

[3] Jodi M. Brown, Patrick A. Langan, and David J. Levin, *Felony Sentences in State Courts,* 1996. Bureau of Justice Statistics (1999).

understandably give priority to crimes with greater penalties. Similarly, sentencing guidelines do not provide for increased penalties for trafficking crimes involving truly large scale trafficking — more than 50 firearms. The most important conclusion to be drawn from this case review is that prosecutors are finding ways to prosecute criminal traffickers despite these issues. Law enforcement authorities, however, must in effect build a double case against many firearms traffickers, identifying non-trafficking conduct that can be the basis for a strong criminal case.

Gun traffickers play a critical and deadly role in the chain of violence. They are a principal source of firearms for criminals. Although some guns are bought legally and used in crime, the many thousands of guns that traffickers supply illegally, without a Brady background check or an FFL transfer record that enables tracing, are firearms that are likely to be associated with other crimes. While the success of the firearms trafficking enforcement effort is to be measured by a reduction in gun violence, it is clear that enforcing Federal laws against firearms traffickers has a deservedly high priority.

1. INTRODUCTION

When the Gun Control Act of 1968 (GCA) was enacted, Congress declared that its primary purpose was to "keep firearms out of the hands of those not legally entitled to possess them . . . and to assist law enforcement authorities in the States and their subdivisions in combating . . . crime in the United States." Many criminals obtain their guns from the illegal market supplied by a variety of sources: unlicensed sellers who buy guns with the purpose of reselling them; fences; corrupt Federal firearms licensees (FFLs); and straw purchasers who buy guns for other unlicensed sellers, criminals, and juveniles.[4] It is the responsibility of the Bureau of Alcohol, Tobacco and Firearms (ATF), often working together with State and local law enforcement agencies, to investigate this criminal trafficking, arrest the perpetrators, and refer them to U.S. Attorneys for prosecution.

ATF has pursued cases against firearms traffickers throughout its history. Starting in 1996, through the Youth Crime Gun Interdiction Initiative and related enforcement and training programs, ATF increased its efforts to work with State and local law enforcement agencies to expand the tracing of guns recovered by police, and to make discovery of the source of the crime gun, and criminal prosecution of the illegal supplier of firearms, a routine aspect of the investigation and prosecution of armed criminals and juveniles.

This report documents ATF criminal investigations involving firearms traffickers that were initiated from July 1996 through December 1998. The report begins by explaining how the nation's firearms laws prohibit illegal transfers of firearms. It then presents an analysis of more than 1,500 ATF firearms trafficking investigations as reported in a survey of ATF field agents. The report also analyzes the agents' reports of the disposition of the cases — the violations referred by ATF, the charges filed and verdicts won by U.S. Attorneys and State and local prosecutors, and the sentences imposed by courts. Finally, the report discusses lessons learned from these investigations and conclusions about the further development of an integrated firearms enforcement strategy against armed criminals and the illegal suppliers who fuel their gun violence.

During the period of review, trafficking investigations comprised only a portion of ATF's firearms investigations. Most ATF firearms investigations specifically targeted violent offenders — armed career criminals, armed narcotics traffickers, and felons in possession of firearms — as part of locally designed strategic efforts to reduce drug trafficking and violent gang activity, and violent crime generally. The gun trafficker, however, also plays a critical role in the chain of violence. Whether a single straw purchaser buying a TEC-9 at a gun show for a high school student, a corrupt licensed dealer selling a silencer to a felon who fails a Brady check, or a gun runner funneling guns to an urban gang, these are violators who demand the full enforcement attention of ATF and our State and local partners. In addition to being traffickers, these violators are also often themselves violent offenders. ATF issues this report to assess and strengthen our efforts, to broaden public understanding, and to support other law enforcement authorities in their efforts to confront the illegal trafficker in guns.

---

[4] Under current law, a "straw purchase" occurs when the actual buyer of a firearm uses another person, the "straw purchaser," to execute the paperwork necessary to purchase a firearm from an FFL. The "straw purchaser" violates the GCA by making a false statement with respect to information required to be kept in the FFL's records.

2. FIREARMS TRAFFICKING AND FIREARMS TRAFFICKING LAWS

The Gun Control Act has a dual framework for keeping firearms out of the wrong hands – prohibiting certain people from possessing guns and regulating the sale of guns. A variety of provisions of the GCA are used to investigate and prosecute firearms traffickers. Unlike criminal misuse of guns, however, firearms trafficking charges can be hard to prove. This has often made cases against firearms traffickers more complicated and challenging than other kinds of gun cases.

2-1. Firearms Trafficking and Diversion

Because initially all crime guns start off as legally owned firearms, the term "firearms trafficking" refers to the illegal diversion of legally owned firearms from lawful commerce into unlawful commerce, often for profit. The term "trafficking" has a different meaning in the firearms context than in the context of drug trafficking, where it usually refers to the illegal manufacture, transportation, and smuggling of large quantities of illicit drugs.

Diversion and Stolen Firearms

ATF uses the term "diversion," in addition to "trafficking." "Diversion" is a broader term than "trafficking," and encompasses any movement of firearms from the legal to the illegal marketplace through an illegal method or for an illegal purpose. For example, a criminal who steals a firearm from a licensee for his own personal use is participating in the illegal diversion of a firearm, but he is not a trafficker. Thus, while the theft of firearms may involve a criminal stealing one or more firearms for his own use, or may involve subsequent trafficking, addressing stolen firearms is an important part of a firearms trafficking strategy because theft constitutes one means of the illegal supply of firearms. In this report, "trafficking" and "diversion" are used synonymously, even though, strictly speaking, diversion is a broader term. Investigations involving gun theft are included in the report only when they also involve trafficking in stolen guns.

Types of Trafficking

Firearms trafficking includes:

- Trafficking in *new* firearms, interstate and intrastate, including by federally licensed firearms dealers, large scale straw purchasers or straw purchasing rings, or small scale straw purchasers, from gun stores, gun shows, or other premises;

- Trafficking in *secondhand* firearms, interstate and intrastate, including by licensed firearms dealers, including pawnbrokers, large scale straw purchasers or straw purchasing rings; small scale straw purchasers, unlicensed sellers, including at gun shows, flea markets, or through newspaper ads, gun magazines, the Internet, and personal associations; and bartering and trading within criminal networks.

- Trafficking in new and secondhand *stolen* firearms, involving guns stolen from federally licensed dealers, including pawnbrokers, manufacturers, wholesalers, and importers, theft from common carriers, home invasions, and vehicle theft.

Firearms theft is often associated with trafficking. ATF investigates thefts from Federal firearms licensees who are required to report thefts or losses to ATF. ATF also investigates thefts from common carriers, some of which voluntarily report thefts to ATF. State and local law enforcement officials may involve ATF when there is a wave of local burglaries apparently for the purpose of obtaining firearms that may then be trafficked. Federal law does not require individual firearms owners to report the theft or loss of a firearm, though owners may report losses to local authorities and/or the FBI in order to assist law enforcement and facilitate the return of the stolen property should it be recovered.

2-2. Laws Prohibiting Trafficking in Firearms

One of the primary purposes of the Gun Control Act of 1968 (GCA) was to provide support to State and local law enforcement officials in their fight against violent crime and to assist them in enforcing their own firearms laws. Prior to enactment of the GCA, differences among State laws made it difficult for State and local officials to enforce their laws. Unrestricted mail-order sales of firearms and the ability of persons to acquire firearms in other less restrictive States prevented many States from effectively enforcing their laws. The GCA remedied this situation by providing a Federal scheme that channels interstate commerce in firearms through Federally licensed dealers, who are required to transfer firearms in accordance with State and local law. The interstate controls of the GCA also generally require Federal firearms licensees to transfer firearms only to residents of the State where their premises are located and restrict unlicensed persons to acquiring firearms within their State of residence. Licensees, however, may transfer long guns to non-residents if they meet in person with the purchaser and the sale complies with the applicable State laws for both buyer and seller.

The GCA also imposes a number of restrictions on persons who acquire or attempt to acquire firearms as well as restrictions on persons who dispose of firearms.

Prohibitions on Firearms Acquisition

The GCA makes it unlawful for certain "prohibited persons," such as convicted felons, fugitives, persons adjudicated mentally ill, and certain domestic violence offenders to possess firearms.[5] With limited exceptions, it is unlawful for anyone under the age of 18 to possess a handgun. Persons who acquire firearms from FFLs must certify their eligibility to purchase firearms, and it is illegal to provide false information in purchasing a firearm. A person purchasing a firearm from an FFL may only acquire a firearm for himself or herself unless he or she is purchasing a firearm for someone else as a gift.

Regulation of Firearms Disposition

Sales regulation is aimed at preserving a State's authority to control commerce in firearms within its borders, as well as to help restrict supply to persons prohibited by Federal law from buying firearms. The GCA requires individuals who are engaged in the business of dealing in firearms to obtain a Federal firearms license. A condition of the Federal license is that the licensee also obeys State and local requirements governing gun sales. The GCA's "interstate controls" make it unlawful for unlicensed individuals to sell any firearms across State lines, or for Federal firearms licensees to deliver handguns to residents of another State. An FFL may not sell a handgun to anyone under the age of 21, or a long gun to anyone under the age of 18. Since 1994, the Brady Act has required FFLs to conduct background checks to determine whether a purchaser is prohibited. FFLs are required to maintain records of the acquisition and disposition of firearms, report multiple handgun sales, report lost or stolen firearms to ATF, and provide transaction records for firearms traces initiated by law enforcement.[6]

2-3. Regulatory and Criminal Enforcement

ATF's firearms trafficking efforts include both regulatory and criminal enforcement of these requirements. Regulatory enforcement aims to ensure FFL compliance with the GCA rules

---

[5] 18 U.S.C. § 922.

[6] Federal law does not require all sellers of guns to obtain a Federal firearms license. In fact, the GCA specifically provides that a person who makes "occasional sales, exchanges, or purchases of firearms for the enhancement of a personal collection or for a hobby, or who sells all or part of his personal collection of firearms" is not required to obtain a firearms license 18 U.S.C. § 922 (a) (21) (c). Unlicensed sellers are prohibited from knowingly selling a firearm to a person prohibited by law from possessing a firearm. However, they are not required to conduct Brady background checks to determine if their buyer is a prohibited person, nor are they required to maintain records that permit the firearm to be traced if it is recovered by law enforcement officials in connection with a crime.

---

governing the conduct of the firearms business, and focuses its inspection efforts on FFLs associated with indicators of trafficking. FFLs that violate the rules may be subject to revocation of their license, as well as to criminal prosecution. Regulatory enforcement is described in a recent ATF report.[7]

2-4. Trafficking Charges and Penalties

No section of the GCA is specifically devoted to punishing the diversion of firearms from lawful to unlawful channels. There are a variety of firearms violations that may be charged in a case involving trafficking. Certain GCA violations, however, are more relevant than others to building cases against traffickers. The charging decision often involves an assessment of the full scope of conduct, the penalties attached to particular charges, and possible defenses at trial.

The two most straightforward trafficking offenses are engaging in the business of dealing in firearms without a license, *see* 18 U.S.C. § 922(a)(1)(A), and traveling into another State to acquire a firearm in furtherance of an intent to violate section 922(a)(1)(A), *see* 18 U.S.C. § 924(n). There are many trafficking cases, however, in which neither of these statutes will be cited, primarily for two reasons. First, cases brought against unlicensed dealers under section 922(a)(1)(A) present prosecution challenges because of the statutory definition of "engaged in the business" in the 1986 amendments to the GCA.[8] Second, ATF firearms trace information indicates that most firearms traffickers acquire firearms in the State in which they are selling them. In such cases, section 924 (n) is not applicable.

When these two offenses are not available, alternative charges are brought. These charges may not, at first glance, appear relevant to firearms trafficking. Depending on the facts of the case, a defendant who was engaged in trafficking may be charged with one or more of the following violations:

- The serial numbers on firearms are often obliterated to make it impossible to trace the firearms. This happens in conjunction with stolen firearms, and it is also an indicator of firearms trafficking. *If it can be established that the defendant knowingly possessed firearms with obliterated serial numbers, and the firearm has at any time been shipped or transported in interstate commerce, then the defendant may be charged with violating 18 U.S.C. § 922(k).*

- Firearms traffickers often present false identification documents to Federal firearms licensees when obtaining firearms. This ensures that the firearms cannot be traced back to the trafficker and circumvents the required Brady background check. In addition, traffickers often use "straw purchasers" to falsely represent to the licensee that they are the actual purchasers of the firearm. *If it can be established that the trafficker or "straw purchaser" lied to the licensee about a material fact, or presented a false identification document in connection with the purchase of the firearm, the trafficker or "straw purchaser" may be charged with violating 18 U.S.C. § 922(a)(6) or aiding and abetting such a violation.* Straw purchasing itself, that is, the practice of buying firearms from an unlicensed dealer on behalf of someone else, is not unlawful.

- Criminals often steal firearms for the purpose of trafficking them. *If it can be established that the firearms were stolen, and that the defendant transported the*

---

[7] *Commerce in Firearms in the United States,* Department of the Treasury, Bureau of Alcohol, Tobacco and Firearms, February 2000.

[8] The term "engaged in the business" means . . . "as applied to a dealer in firearms, . . . a person who devotes time, attention, and labor to dealing in firearms as a regular course of trade or business with the principal objective of livelihood and profit through the repetitive purchase and resale of firearms, but such term shall not include a person who makes occasional sales, exchanges, or purchases of firearms for the enhancement of a personal collection or for a hobby, or who sells all or part of his personal collection of firearms. . . ." 18 U.S.C. § 921(a)(21)(C).

firearms in interstate commerce knowing or having reasonable cause to believe that they were stolen, the defendant may be charged with violating 18 U.S.C. § 922(i). If it can be established that the defendant stole the firearms from a licensee and the firearm has been shipped in interstate or foreign commerce, the defendant may be charged with violating 18 U.S.C. § 922(u).

- Many traffickers distribute firearms to criminals and gang members knowing that the recipients intend to use the firearms for criminal purposes. *If it can be established that the defendant transferred the firearm, knowing that the firearm will be used in a crime of violence or a drug trafficking crime, the defendant may be charged with violating 18 U.S.C. § 924(h).*

- Traffickers may distribute handguns to juvenile gang members and other individuals under the age of 18. *If it can be established that the defendant transferred a handgun to an individual knowing or having reasonable cause to believe that the transferee was under the age of 18, and none of the statutory exemptions apply, the defendant may be charged with violating 18 U.S.C. § 922(x)(1).*[9]

- Certain "corrupt" dealers cooperate with traffickers by knowingly maintaining false records of acquisition and disposition, or failing to keep required records. *If it can be established that a firearms licensee knowingly made a false statement in required records, the licensee may be charged with violating 18 U.S.C. § 924(a)(1)(A).*

- *A person who illegally traffics in certain weapons covered by the National Firearms Act (NFA), such as machine guns, short barrel shotguns, short barrel rifles, and silencers, can be prosecuted for violations of the NFA, 26 U.S.C. Chapter 53.*

- Finally, in some cases the individual suspected of trafficking is also a felon. While it may be difficult to prove that the defendant was engaged in the business of dealing in firearms, the evidence clearly establishes that he was a felon in possession of firearms. *If it can be established that the defendant possessed firearms after being convicted of a crime punishable by a term exceeding one year, the defendant may be charged with violating 18 U.S.C. § 922(g)(1). If it can be established that any person transferred a firearm to a prohibited person knowing or having reasonable cause to believe that the person was prohibited then the transferor may be charged with violating 18 U.S.C. 922(d). If the defendant had been convicted of three prior violent felonies or serious drug offenses, however, the defendant may receive an enhanced 15 year sentence under the Armed Career Criminal Act. 18 U.S.C. § 924(e).*

These examples are not exhaustive. Rather they illustrate the principle that not all trafficking cases involve criminal statutes readily identified as penalizing "trafficking" or involving "trafficking" charges. Although these charges are not uniquely "trafficking" charges, they are the statutory tools that investigators and prosecutors most often use in trying to punish and deter firearms traffickers. In ATF's view, even though the perpetrator may not be charged with "trafficking in firearms," as there is no such specific criminal violation, the goal is to get illegal traffickers off the street and prevent others from emerging, by prosecuting them to the fullest extent possible under the Federal firearms laws.

One of the conclusions drawn from the case review presented here is that persons who traffick in firearms are often not being prosecuted for that conduct; rather, they are instead being prosecuted for other related conduct.

---

[9] Exceptions from the transfer and possession prohibitions of § 922 (x) are provided for the following temporary transfers, if the juvenile has prior written consent from a parent and the consent is in the juvenile's possession at all times: (1) employment; (2) ranching or farming; (3) target practice; (4) hunting; and (5) firearms safety instruction. Permanent transfers of handguns to juveniles are also authorized under the statute for juveniles who are members of the Armed Forces.

3. REVIEW OF FIREARMS TRAFFICKING INVESTIGATIONS
July 1996 - December 1998

The case review presented in this report originated in a Congressional request for information about ATF's enforcement activities. Specifically, the House and Senate Committees on Appropriations requested ATF to report on trafficking investigations by February 1999 in connection with funding for ATF's Youth Crime Gun Interdiction Initiative (YCGII), the component of ATF's firearms enforcement programs focused on illegal acquisition, possession, and use of guns by youth and juveniles.[10]

In response, ATF Headquarters requested all ATF Special Agents in Charge to provide information on all firearms trafficking investigations in their respective areas between July 1996 (the commencement date of YCGII) and December 1998 (the end of the last calendar year before February 1999). A survey was sent to each Field Division requesting information for each investigation (see Appendix B).

The 23 ATF Field Divisions submitted a total of 1,530 reports on investigations, including ongoing investigations and perfected cases referred for prosecution. Information on 648 investigations involving youth and juveniles were reviewed and provided the basis for a report to Congress on the performance of YCGII in February 1999.[11] In this report, ATF and an outside researcher review all 1,530 investigations, using the survey forms submitted.[12]

This report also reviews the disposition of cases referred by ATF for prosecution. To develop disposition information, ATF in December 1999 sent supplementary surveys for the 1,530 submitted investigations to the 23 ATF Field Divisions (see Appendix B). All surveys were returned to ATF Headquarters by March 15, 2000 for analysis by ATF personnel and outside researchers. Case disposition information was reviewed by outside researchers working with the Bureau of Justice Statistics, which has statutory authority for collecting and maintaining Federal case disposition information.[13]

The case review is presented in five sections: initiation of firearms trafficking investigations; traffickers and trafficking channels; firearms misuse, felons, and trafficking investigations; characteristics of the investigations; and characteristics of case dispositions.

---

[10] The Statement of Managers accompanying the 1998 Conference Report stated that: "the conferees believe that the proposed increase in funding must be supported by evidence of a significant reduction in youth crime, gun trafficking, and gun availability. The conferees would like to see additional evidence linking the Youth Crime Gun Interdiction Initiative (YCGII) to a corresponding decrease in gun trafficking among youths and minors. Therefore, the conferees direct ATF to report no later than February 1, 1999 on the performance of YCGII." Conference Report to Accompany H.R. 4328, October 19, 1998.

[11] See *Youth Crime Gun Interdiction Initiative: Performance Report.* Report to the Senate and House Committees on Appropriations Pursuant to Conference Report 105-825, October 1998. Department of the Treasury, Bureau of Alcohol, Tobacco, and Firearms, 1999.

[12] Dr. Anthony A. Braga of the John F. Kennedy School of Government, Harvard University.

[13] Dr. Anthony A. Braga of the John F. Kennedy School of Government, Harvard University and Dr. Joel Garner of the Joint Centers for Justice Studies, Sheperdstown, West Virginia.

3-1. Initiation of Firearms Trafficking Investigations

Traditional methods. ATF firearms trafficking investigations were most frequently initiated after a referral from other law enforcement agencies or through information provided by confidential informants.

FFL cooperation. In about nine percent of the investigations, Federal firearms licensees (FFLs) reported suspicious activity to ATF, such as suspected straw purchasing or the use of false identification to obtain firearms. Almost eight percent of the investigations were initiated by FFLs reporting burglaries, thefts, and robberies to ATF. ATF maintains a database on firearms that FFLs report as stolen or missing; this information is used to initiate investigations and in connection with regulatory compliance inspections.

Innovative use of information resources: tracing and multiple sales forms. Although many investigations were initiated via traditional law enforcement leads, many others were initiated through the use of crime gun trace information and multiple sales information.[14] Nearly 20 percent of the investigations were initiated through trace analysis after the recovery of crime guns, and slightly more than 13 percent of the investigations were initiated after review-

ing multiple firearms sales records. Almost 30 percent of all ATF firearms trafficking investigations between July 1996 and December 1998 were initiated either through the analysis of trace data, multiple firearms sales records, or both (448 of 1,530). Beyond the initiation of investigations, tracing was used as an investigative tool to gain information on recovered crime guns in 60 percent of the investigations.[15]

These findings show that crime gun trace analysis and Project LEAD, the National Tracing Center's firearms trafficking information system, are being widely used to increase ATF's productivity in the development of firearms trafficking investigations. With Project LEAD available on-line to ATF field offices and State and local task forces working with ATF field offices since November 1999, and assuming continuing increased participation in tracing by State and local law enforcement agencies, trace and multiple sales information can be expected to grow in importance as a source of leads to traffickers. In addition, as law enforcement officials develop the ability to connect spent cartridges and bullets to firearms through the National Integrated Ballistics Information Network (NIBIN), the ability to identify illegal sources of firearms will continue to grow.

---

[14] 18 U.S.C. § 923 (g) (3) (B). In November 1998, the National Tracing Center began entering multiple sales information into Project LEAD, ATF's firearms trafficking information system, making it available for trafficking investigations.

[15] Information obtained from the Supplemental Survey.

Table 1. Initiation of ATF Firearms Trafficking Investigations

Reason	Number of investigations	Percent
Referral from another State, local, or Federal agency	409	26.7
Confidential informant	352	23.0
Trace analysis after firearms recovery	296	19.4
Review of multiple sales forms	205	13.4
FFL reported suspicious activity	139	9.1
Developed from another ATF investigation	127	8.3
FFL reports burglary/ theft / robbery to ATF	115	7.5
ATF initiated investigation of suspicious activity (e.g. gun show task force, interstate theft task force, etc.)	81	5.3
ATF Regulatory inspection of FFL records	43	2.8
Tip by concerned citizen or anonymous source	37	2.4
Other	9	0.6

Number of investigations included = 1,530

Note: Sum may exceed 100 percent since investigations may be included in more than one category.

3-2. Traffickers and Trafficking Channels

For law enforcement agencies in any community, a key challenge is to identify the sources of illegal supply of firearms, the types of traffickers, and the predominant types of trafficking channels. For instance, if criminals in a particular community are obtaining firearms through theft from lawful firearms owners, law enforcement and community public safety strategies must focus on reducing theft.[16] To the extent that the illegal market in firearms is supplied by transactions at FFLs in new and secondhand firearms, whether pawnbrokers, other retail dealers, or residential licensees, an enforcement response to these sources is required.

The records of purchase denials under the Brady Act show that prohibited persons, including some violent felons, do seek to acquire guns from FFLs.[17] This review confirms that retail transactions in new and secondhand firearms with FFLs and unlicensed sellers supply the illegal market, and also that stolen firearms are trafficked by unlicensed dealers.

Firearms Trafficking Channels Identified in ATF Investigations

Retail transactions: straw purchasing, unlicensed sellers, and corrupt FFLs. The most frequent type of trafficking channel identified in ATF investigations is straw purchasing from federally licensed firearms dealers. Nearly 50 percent of the ATF investigations involved firearms being trafficked by straw purchasers either directly or indirectly. The investigations also involve trafficking by unlicensed sellers (more than 20 percent); by federally licensed dealers (just under 9 percent); and diversion from gun shows and flea markets by FFLs or unlicensed sellers (about 14 percent).

Theft as a source of trafficked firearms. Firearms stolen from FFLs, residences, and common carriers were involved in 26 percent of the trafficking investigations. ATF's trafficking investigations show that firearms may be stolen and subsequently trafficked from a variety of sources, including FFL dealers, pawnbrokers, residences, and common carriers. These stolen firearms may be sold subsequently by individuals and groups specializing in firearms trafficking or by those fencing a variety of stolen goods.[18]

---

[16] See e.g. James D. Wright, 1995, "Ten Essential Observations on Guns in America," *Society* March/April: 63 – 68; Philip J. Cook, Stephanie Molliconi, and Thomas B. Cole, 1995, "Regulating Gun Markets," *Journal of Criminal Law and Criminology* 86: 59-92; James D. Wright and Peter H. Rossi, 1994, *Armed and Considered Dangerous: A Survey of Felons and Their Firearms*, Expanded Edition, New York: Aldine De Gruyter; Bureau of Justice Statistics. 1993. "Survey of State Prison Inmates, 1991" Washington, DC: US Department of Justice; Joseph F. Sheley and James D. Wright. 1993. "Gun Acquisition and Possession in Selected Juvenile Samples." *Research in Brief*. Washington, DC: US Department of Justice.

[17] Since the passage of the Brady Act, more than 500,000 prohibited persons have been prevented from buying firearms from FFLs.

[18] Two studies conducted in New York suggested that firearms trafficking was an important source of guns for criminals. A review of a 1973 analysis of stolen firearms by the New York City Police Department showed that, beyond residential burglaries, firearms are stolen from a variety of sources, including burglaries and robberies of manufacturers, licensed dealers, and during shipping, that stolen firearms are sometimes subsequently trafficked by thieves, and that stolen firearms are rapidly recovered in crime after theft. Steven Brill, 1977, *Firearms Abuse: A Research and Policy Report*, Washington, D.C.: Police Foundation. About 40 percent of 22 stolen firearms examined by Brill were recovered in crime within six months after the theft; Brill rightly cautions against over-interpreting these data due to small sample size. A 1992 examination of illegal trafficking of firearms into New York City noted that straw purchasers and corrupt federally licensed dealers illegally divert firearms to criminal consumers. Jeremy Travis and William Smarrito. 1992, "A Modest Proposal to End Gun Running in America," *Fordham Urban Law Journal* 19: 795 – 811.

Regional variations. Regional variations in the trafficking channels in ATF investigations suggest that the illegal market in guns may operate differently in different areas of the country. For instance, straw purchasing was involved in almost two thirds of ATF's trafficking investigations in its Northeast region, but closer to a quarter of its trafficking investigations in its Southwest and Western regions.[19] However, without additional information, it is not possible to know whether regional differences reflect differences in investigative practice, in the illegal market, or both.

Table 2. Sources of Firearms Trafficking Identified in ATF Investigations

Source	Number of investigations	Percent
Firearms trafficked by straw purchaser or straw purchasing ring	709	46.3
Trafficking in firearms by unlicensed sellers*	314	20.5
Trafficking in firearms at gun shows and flea markets	212	13.9
Trafficking in firearms stolen from FFL	209	13.7
Trafficking in firearms stolen from residence	158	10.3
Firearms trafficked by licensed dealer, including pawnbroker	133	8.7
Street criminals buying and selling firearms from unknown sources**	95	6.2
Trafficking in firearms stolen from common carrier	31	2.0
Unlicensed manufacture of common firearms or NFA weapons	16	1.0
Other sources (e.g. selling firearms over Internet, illegal pawning)	18	1.1

Number of investigations included = 1,530

Note: Sum may exceed 100 percent since investigations may be included in more than one category.

 * As distinct from straw purchasers and other traffickers.

** These were investigations in the early stages where trafficking channels were not yet fully clear.

---

[19] The Northeast region consisted of trafficking investigations submitted by the Boston, New York, Philadelphia, Baltimore, and Washington, D.C. Field Divisions. The Southeast region consisted of trafficking investigations submitted by the Charlotte, Miami, Tampa, Atlanta, Nashville, and New Orleans Field Divisions. The Central region consisted of trafficking investigations submitted by the Chicago, Columbus, Detroit, Kansas City, Louisville, and St. Paul Field Divisions. The Southwest region consisted of trafficking investigations submitted by the Phoenix, Dallas, and Houston Field Divisions. The Western region consisted of trafficking investigations submitted by the Los Angeles, San Francisco, and Seattle Field Divisions.

Volume of Firearms Diverted, By Trafficking Channel

The types of trafficking channels differ in the mean number of guns per investigation and the overall number of guns associated with them.

Trafficking by corrupt FFLs. Licensed dealers, including pawnbrokers, have access to a large volume of firearms, so a corrupt licensed dealer can illegally divert large numbers of firearms. Although FFL traffickers were involved in the smallest proportion of ATF trafficking investigations, under 10 percent, FFL traffickers were associated with by far the highest mean number of illegally diverted firearms per investigation, over 350, and the largest total number of illegally diverted firearms, as compared to the other trafficking channels.

Gun shows. Investigations involving gun shows involved the second highest number of trafficked guns per investigation, over 130, and were associated with over 26,000 illegally diverted firearms.

Straw purchasers and unlicensed sellers. Straw purchasing rings and small scale straw purchasers averaged about half as many illegally diverted firearms per investigation (37) as unlicensed dealers (75), but the two types of trafficking channels were associated with a similar number of trafficked firearms overall, over 20,000.

Stolen firearms. Investigations involving firearms stolen from residences and FFLs were associated with the smallest mean number of guns per investigation. Because of the small number of investigations involving thefts of firearms from common carriers, this trafficking channel yielded the smallest total number of firearms, although it averaged a substantial number of illegally diverted firearms per investigation.[20]

---

[20] Common carriers are not required to report thefts to ATF, so relatively few such investigations are initiated. However, ATF will soon propose a regulation that would require FFLs to report firearms lost in shipment.

Table 3. Volume of Firearms Diverted, By Trafficking Channel

Source	Number of		Average number of firearms per invest.*	Number of firearms in more than half the investigations**
	investigations	firearms		
Firearms trafficked by straw purchaser or straw purchasing ring	695	25,741	37.0	14
Trafficking in firearms by unlicensed sellers***	301	22,508	74.8	10
Trafficking in firearms at gun shows and flea markets	198	25,862	130.6	40
Trafficking in firearms stolen from FFLs	209	6,084	29.1	18
Trafficking in firearms stolen from residence	154	3,306	21.5	7
Firearms trafficked by licensed dealer, including pawnbroker	114	40,365	354.1	42
Trafficking in firearms stolen from common carrier	31	2,062	66.5	16

Number of investigations included = 1,470

Number of firearms included = 84,128

Note: Sum may exceed 100 percent since investigations may be included in more than one category. Excludes 60 investigations with an unknown number of trafficked firearms.

 * Averages were calculated using the mean.

** This measure used the median.

*** As distinct from straw purchasers and other traffickers.

Federally Licensed Dealers Conspiring With Other Traffickers

In most firearms trafficking investigations, ATF agents uncovered only one pathway through which guns were diverted or trafficked (80.8 percent, 1,237 of 1,530). Firearms trafficking investigations involving FFLs often involve a mixture of channels, including gun shows (31 percent), straw purchasers (28 percent), and unlicensed sellers (12 percent).

Table 4. Other Trafficking Channels Involved in FFL Trafficking Investigations

Trafficking channels	Number	Percent
Number of investigations involving FFLs	133	100.0
FFL trafficking alone	48	36.1
Multiple channels	85	63.0
Multiple channels (may be more than one)		
Trafficking in firearms at gun shows and flea markets	41	30.8
Firearms trafficked by straw purchaser or straw purchasing ring	37	27.8
Trafficking in firearms by unlicensed sellers*	16	12.0
Trafficking in firearms stolen from residence	6	4.5
Trafficking in firearms stolen from FFLs	3	2.3

Note: Sum may exceed 100 percent since investigations may be included in more than one category.

* As distinct from straw purchasers and other traffickers.

Federally Licensed Dealers' Impact on the Volume of Illegal Supply

Licensed firearms dealers' access to large volumes of firearms can influence the number of firearms illegally diverted in investigations involving gun shows, straw purchasers, and unlicensed sellers. An unlicensed dealer conspiring with a corrupt FFL, for example, may be able to divert larger numbers of firearms to prohibited persons than an unlicensed seller acting alone. When corrupt FFLs act in conjunction with other types of traffickers, the average number of guns diverted per investigation increases dramatically when compared to investigations of gun shows, straw purchasers, and unlicensed sellers that do not involve conspiracy with a corrupt FFL.

Table 5. The Influence of FFL Traffickers on the Number of Firearms Trafficked by Straw Purchasers, Unlicensed Sellers, and at Gun Shows

Source	No FFL involvement		FFL involvement	
	Number of investigations	Average number of firearms per investigation*	Number of investigations	Average number of firearms per investigation*
Firearms trafficked by straw purchaser or straw purchasing ring	659	32.8	36	114.8
Trafficking in firearms by unlicensed sellers**	285	46.6	16	576.6
Trafficking in firearms at gun shows and flea markets	161	87.9	37	316.5
Firearms trafficked by licensed dealer, including pawnbroker; no other trafficking channels	——	——	48	562.6

Number of investigations included = 1,470.

Note: Sum may exceed 100 percent since investigations may be included in more than one category. Excludes 60 cases with an unknown number of trafficked firearms.

 * Averages were calculated using the mean.
** As distinct from straw purchasers and other traffickers.

The Business Premises of Licensed Dealers in Trafficking Investigations

Enforcement information provides some insight into the relative roles of various elements of the licensed dealer population. A 1998 random sample of FFLs identified 69 percent of licensees as retail gun dealers and 10 percent as pawnbrokers. The remaining 21 percent were collectors of curios and relics, manufacturers of firearms and ammunition, and importers. Of the retail dealers and pawnbrokers, 44 percent operated from commercial premises (a quarter of these were gun shops or sporting goods and hardware outlets, while the rest were businesses not normally associated with a gun business, such as funeral homes and auto parts stores), and 56 percent from residential premises, down from 74 percent in 1992.[21]

A 20 percent sample (26 of 133) of FFL trafficking investigations under review here were randomly selected to examine the type of business premises maintained by the FFLs under investigation.

Slightly more than 38 percent were retail dealers and exactly the same number were pawn shops; 23 percent were "kitchen table" or residential dealers. Half of the licensed dealers who maintained a gun store business premises also sold firearms at gun shows. One third of the residential dealers and 20 percent of pawnbrokers also sold firearms at gun shows.

Because of the relatively small sample of cases examined, this table shows the variety of business premises of licensed dealers engaged in trafficking, but cannot be used as a basis for strong inferences about the relative importance of one type of business premise relative to another in firearms trafficking investigations.

Table 6. Business Premises of a 20 Percent Random Sample of FFLs Involved in Trafficking Investigations

FFL business premises	Number of investigations	Percent of all investigations	Percent of category selling at gun shows
Total	26	100.0	35.0
Gun store business premises	10	38.5	50.0
Pawn shop business premises	10	38.5	20.0
Residential business premises	6	23.0	33.3

Note: Since this is a 20 percent sample, it may not be representative of all FFLs.

---

[21] *Commerce in Firearms in the United States*, February 2000, Department of the Treasury, Bureau of Alcohol, Tobacco and Firearms.

Gun Shows and the Diversion of Firearms

Gun shows are a special case of the retail sale of firearms. Unlike at other venues, both federally licensed dealers and unlicensed sellers sell guns at gun shows. While federally licensed firearms dealers perform background checks, unlicensed sellers are not required to do so. Gun shows are also places where buyers can choose to buy from the primary (firearms sold by FFLs) or secondary (firearms resold by unlicensed sellers) firearms markets. Secondhand firearms are far more difficult than new guns for law enforcement officials to trace to the most recent seller. This is because secondhand firearms likely have left the hands of FFLs, who are required to keep records, into the hands of unlicensed persons who are not required to keep records. Even if the secondhand guns are resold to an FFL, they are untraceable, because the trace will effectively end at the last sale in the unbroken chain of licensed sellers. The access to anonymous sales and large numbers of secondhand firearms makes gun shows attractive to criminals.

In the ATF trafficking investigations reviewed here, gun shows were associated with the diversion of approximately 26,000 firearms. Trafficking of firearms at and through gun shows was more prevalent in the trafficking investigations submitted by the Southwest (22 percent) and Western (21 percent) regions when compared to trafficking investigations submitted by the Central (15 percent), Southeast (12 percent), and Northeast (8 percent) regions of the United States.

A prior review of ATF gun show investigations shows that prohibited persons, such as convicted felons and juveniles, do personally buy firearms at gun shows and gun shows are sources of firearms that are trafficked to such prohibited persons. The gun show review found that firearms were diverted at and through gun shows by straw purchasers, unregulated private sellers, and licensed dealers.[22] Felons were associated with selling or purchasing firearms in 46 percent of the gun show investigations.[23] Firearms that were illegally diverted at or through gun shows were recovered in subsequent crimes, including homicide and robbery, in more than a third of the gun show investigations.[24]

---

[22] *Gun Shows: Brady Checks and Crime Gun Traces,* Department of the Treasury and Department of Justice, January 1999: Table 3.

[23] *Gun Shows: Brady Checks and Crime Gun Traces,* Department of the Treasury and Department of Justice, January 1999: Table 3.

[24] *Gun Shows: Brady Checks and Crime Gun Traces,* Department of the Treasury and Department of Justice, January 1999: Table 4.

Straw Purchasers and the Diversion of Firearms

Investigations involving straw purchasers averaged 37 firearms per investigation, but because of the large number of investigations involving straw purchasing (695, 46 percent of all the trafficking investigations), this trafficking channel was associated with nearly 26,000 illegally diverted firearms. The 550 investigations involving straw purchasers without any additional trafficking channels averaged 26 illegally diverted firearms per investigation, and a total of just over 14,000 illegally diverted firearms. Although the average number of firearms trafficked per investigation is relatively small when compared to other trafficking channels, illegally diverted firearms associated with straw purchasers represent nearly a third of the illegally diverted firearms in all ATF investigations initiated between July 1996 and August 1998. Thus, despite the relatively small numbers per investigation, straw purchasers represent a significant overall crime and public safety problem.

Straw purchasers may be the instruments of criminals or traffickers who obtain the straw purchaser's services, or they may be unlicensed dealers who set out to use their non-prohibited status to traffic to other persons for a profit. In other words, the straw purchaser may also be the trafficker. In 25 percent of the investigations (387 of 1,530), the straw purchasers were working for traffickers, and in 19 percent of the investigations (292) the straw purchasers were the traffickers.[25] Straw purchasers were often friends (45 percent), relatives (23 percent), and spouses or girlfriends (18 percent) of the firearms traffickers. Almost 28 percent of the investigations involved a business relationship where the trafficker paid the straw purchaser money or drugs to buy firearms. Five percent of the investigations involved a straw purchaser who was a member of the same street gang as the trafficker.

Table 7. Relationships Between Straw Purchaser and Trafficker

Relationship to trafficker	Number of investigations	Percent
Friend	173	44.7
Business (paid with money or drugs to buy firearms)	107	27.6
Relative	89	23.0
Intimate (spouse or girlfriend)	68	17.6
Member of same gang	21	5.4

Number of investigations included = 387

Note: Sum may exceed 100 percent since investigations may be included in more than one category.

---

[25] Table 10, Firearms Traffickers and Felons Identified in Trafficking Investigations.

The Involvement of Juveniles and Youth in ATF Trafficking Investigations

Youth (ages 18 – 24) and juveniles (ages 17 and under) were involved in 42 percent of the ATF firearms trafficking investigations (648 of 1,530) during this period. ATF published an analysis of these investigations in the Youth Crime Gun Interdiction *Performance Report* in February 1999.

Juveniles were primarily involved in the trafficking investigations as possessors of firearms, but they were also involved in the illegal diversion of firearms as firearms traffickers, thieves or robbers, and straw purchasers of firearms.[26] A quarter of the youth and juvenile investigations involved juveniles obtaining firearms through thievery or robbery, and nearly a fifth of these investigations involved juveniles as gun traffickers.[27]

Youth were nearly twice as likely as juveniles to be involved as traffickers, and because of age restrictions on the purchase of firearms, about ten times as likely to be involved as straw purchasers as juveniles.[28] It has been shown that of crime guns recovered and traced by law enforcement officials in 27 cities, more were possessed by 19 year olds than any other age group; crime guns recovered from 18 year olds ranked second.[29]

Table 8: The Role of Youth and Juveniles in Trafficking Investigations

Role of Youth	Number of investigations	Percent
Possessor	337	53.9
Trafficker	236	37.8
Straw purchaser	149	23.8
Thief /robber of firearms	122	19.5

Number of investigations involving at least one youth = 625

Role of Juveniles	Number of investigations	Percent
Possessor	155	74.2
Thief /robber of firearms	53	25.3
Trafficker	40	19.1
Straw purchaser	4	1.9

Number of investigations involving at least one juvenile = 209

---

[26] *Performance Report* for the Senate and House Committee on Appropriations, Youth Crime Gun Interdiction Initiative, Department of the Treasury, Bureau of Alcohol, Tobacco and Firearms, February 1999. pg. 6.

[27] *Performance Report* for the Senate and House Committee on Appropriations, Youth Crime Gun Interdiction Initiative, Department of the Treasury, Bureau of Alcohol, Tobacco and Firearms, February 1999: Appendix, Table 1.

[28] *Performance Report* for the Senate and House Committee on Appropriations, Youth Crime Gun Interdiction Initiative, Department of the Treasury, Bureau of Alcohol, Tobacco and Firearms, February 1999: Appendix, Table 1.

[29] *Gun Crime in the Age Group 18-20*, the Department of the Treasury and the Department of Justice, June 1999.

3-3. Firearms Misuse, Felons, and Trafficking Investigations

Persons prohibited from possessing firearms, including felons, are obtaining guns from the illegal market. While a case review does not measure the full extent of the use of trafficked firearms by prohibited persons or in subsequent crimes, ATF's trafficking investigations show that trafficked firearms are diverted to prohibited persons and are subsequently used in serious crimes, and that some felons are heavily involved in firearms trafficking. Thus, investigations that "follow the crime gun" to its illegal source are not only an effective strategy to disrupt and reduce firearms trafficking in a community, they also are an effective means to apprehend felons, armed career criminals, and narcotics traffickers who possess and misuse firearms.

Trafficked Firearms Subsequently Recovered in Crime

Half of the trafficking investigations involved at least one firearm recovered in crime. More than 14 percent of the trafficking investigations were associated with juvenile possession cases; about 17 percent were associated with homicide and robbery cases, respectively; about 25 percent of the investigations were associated with assault cases; more than 40 percent involved firearms associated with felon in possession cases; and more than 50 percent of these investigations involved firearms associated with illegal possession cases.

Table 9. Known Criminal Uses of Trafficked Firearms

50.4% of the investigations (771 of 1,530) had at least one diverted firearm recovered in a crime.

Crime	Number of cases with at least one firearm recovered in a crime	Percent
Criminal possession (not felon in possession)	390	50.6
Felon in possession	311	40.3
Drug offense	212	27.5
Assault	194	25.2
Homicide	134	17.4
Robbery	127	16.5
Property crime	116	15.0
Juvenile possession	109	14.1
Sexual assault/ rape	15	1.9
Other crime	42	5.4

Number of investigations included = 771

Note: Sum may exceed 100 percent since investigations may be included in more than one category.

The Involvement of Felons in Firearms Trafficking

This review shows that convicted felons are heavily involved in firearms trafficking. Felons were involved in 45 percent (690 of 1,530) of the ATF firearms trafficking investigations in various roles. One third of the thieves and robbers involved in these investigations had at least one prior felony conviction, as did nine percent of straw purchasers working for traffickers, and 24 percent of straw purchasers who were the sole trafficker, five percent of the FFL traffickers, and more than a third of other firearms traffickers, such as unlicensed dealers.

Altogether, there were 2,670 traffickers identified by ATF agents, of whom 25 percent (669) were convicted felons. Felons may be identified as straw purchasers for a number of reasons: an FFL may have a felon as an employee; a felon may be buying from an unlicensed seller who does not conduct Brady checks; a felon may have stolen identity papers; and before the permanent provisions of the Brady Act took effect in November 1998, no criminal history checks were conducted for the purchase of long guns.

Table 10. Firearms Traffickers and Felons Identified in Trafficking Investigations

Role	Total number of traffickers	Number of traffickers with felony convictions	Percent of traffickers with felony convictions
Firearms thief/ robber	616	205	33.3
Straw purchaser working for trafficker	610	53	8.9
Trafficker is the straw purchaser	338	79	23.3
FFL trafficker*	131	7	5.3
Former FFL	25	0	0.0
Other trafficker	950	325	34.2
Total	2,670	669	25.0

Number of investigations included = 1,530

* May include employees of FFL.

3-4. Characteristics of the Investigations

Most ATF trafficking investigations, 68 percent (1,037 of 1,530), involved the cooperation of State and/or local law enforcement agencies. Table 11 shows that about 70 percent of the investigations involved intrastate trafficking, suggesting that where there are applicable State firearms laws, State and local law enforcement have an important role to play in curbing the illegal market in firearms. Table 12 shows that most, but not all, gun trafficking investigations involve relatively small numbers of firearms. Table 13 highlights the relative roles of new, secondhand, and stolen firearms in trafficking investigations. The chain of ownership and status of a firearm can affect law enforcement's ability to identify its illegal supplier. Because of limited recordkeeping and reporting requirements applicable to stolen and secondhand firearms, the illegal sources of new crime guns are much easier to investigate.

Interstate and Intrastate Destinations of Trafficked Firearms

ATF trace analysis reports have documented that the State in which a community is located is generally the largest single source of its traced crime guns.[30] The reports also show that some jurisdictions act as "source" areas (e.g. Florida and Georgia) of crime guns for other "market" areas (e.g. New York and Boston).[31] The review of investigations shows that firearms traffickers engage in intrastate trafficking, interstate trafficking, and international trafficking. Intrastate firearms trafficking was involved in 70 percent of the investigations, while slightly less than half of the investigations involved interstate firearms trafficking. Twenty percent of the investigations involved firearms being trafficked both interstate and intrastate. Firearms were trafficked internationally in about 11 percent of the investigations.

Table 11. Interstate, Intrastate, and International Trafficking in ATF Investigations

Destination of trafficked firearms	Number	Percent
Intrastate	1,072	70.1
Interstate	678	44.3
International (export)	170	11.1
Unknown	53	3.5
Mutually exclusive categories:		
Total	1,530	100.0
Intrastate only	696	45.5
Interstate and intrastate	312	20.4
Interstate only	299	19.5
International only	87	5.7
Interstate, intrastate, and international	47	3.1
Intrastate and international	19	1.2
Interstate and international	17	1.1
Unknown	53	3.5

---

[30] *Youth Crime Gun Interdiction Initiative Trace Analysis Reports, 27 Communities*, Department of the Treasury, Bureau of Alcohol, Tobacco and Firearms, February 1999.

[31] *Youth Crime Gun Interdiction Initiative Trace Analysis Reports, 27 Communities*, Department of the Treasury, Bureau of Alcohol, Tobacco and Firearms, February 1999.

Number of Firearms Trafficked in ATF Investigations

Firearms trafficking mainly involves relatively smaller numbers; 43 percent of the investigations involved 10 firearms or less. Trafficking in large numbers of firearms, however, does occur; the two largest numbers of firearms reported in connection with a single investigation were 10,000 and 11,000, respectively.

Table 12. *Number of Firearms Involved in ATF Trafficking Investigations*

Range	Number of investigtions	Percent
Total number of investigations	1,530	100.0
Less than 5	354	23.1
5 – 10	308	20.1
11 – 20	279	18.2
21 – 50	286	18.7
51 – 100	139	9.1
101 – 250	67	4.4
251 or greater	37	2.4
Unknown	60	3.9

New, Secondhand, and Stolen Firearms

ATF's trace analysis in 27 cities suggests that up to 43 percent of traced firearms are new guns that have moved rapidly from the shelf of an FFL to recovery by law enforcement in three years or less, and, therefore, may have been trafficked.[32] The mere rapid movement of a firearm does not confirm that trafficking has occurred, because a gun may have been purchased by the criminal user or stolen by the criminal user from an FFL or from the person who bought the firearm. ATF, therefore, treats rapid time-to-crime as a trafficking indicator to be investigated with other information.

Focusing exclusively on new guns likely underestimates the true extent of gun trafficking. A National Tracing Center trace usually ends after the first retail sale. Following the trail of the gun further often requires a substantial commitment of investigative effort by the tracing law enforcement agency, a commitment that is often not feasible. ATF's trace analysis reports estimated that between 57 and 71 percent of all traceable crime guns in 27 cities were recovered more than three years after the first retail sale. While a trace of a crime gun may reveal that it was first sold at retail ten years before its recovery in crime, it is nevertheless possible that it was trafficked. The firearm during this time period could be resold to and by FFLs, or by unlicensed sellers on the secondary market, as "used" or "secondhand," and subsequently trafficked.

This review shows that investigations involved the trafficking of new, secondhand, and stolen firearms. Thus, while trace analysis alone cannot shed light on the percentage of older crime guns that are trafficked, ATF investigative experience suggests that trafficking in secondhand firearms is indeed a significant crime and public safety problem. While new firearms figured more prominently, new and secondhand guns were trafficked together in about one third of the investigations.

As documented in Table 3, firearms are stolen from a variety of sources. Depending on the type of theft involved, stolen firearms may range from new to quite old. For example, a burglary of a licensed dealer may yield a cache of new and secondhand firearms while a residential burglary or series of home invasions may yield only older firearms. Both new and secondhand firearms were stolen and subsequently trafficked, with secondhand firearms figuring more prominently among stolen, trafficked firearms.

Table 13. New, Secondhand, and Stolen Firearms in ATF Trafficking Investigations

Type of firearm	Number of investigations	Percent
New firearms	1,120	73.2
Secondhand firearms	899	58.8
Stolen guns	381	24.9
Unknown	27	1.8
Stolen firearms:		
Involves at least one new firearm	223	58.5
Involves at least one secondhand firearm	304	79.8
Mutually exclusive categories for new and secondhand firearms:		
Total	1,530	100.0
New firearms only	604	39.5
New and secondhand firearms	516	33.7
Secondhand firearms only	383	25.0
Unknown	27	1.8

Note: Sum may exceed 100 percent since investigations may be included in more than one category.

---

[32] *Youth Crime Gun Interdiction Initiative Trace Analysis Reports, 27 Communities,* Department of the Treasury, Bureau of Alcohol, Tobacco and Firearms, February 1999. pg. A-3.

3-5. Characteristics of Case Dispositions

An examination of the characteristics of case dispositions shows two things: first, that ATF is vigorously enforcing laws against firearms traffickers with the support of Federal and State prosecutors; and second, that persons who traffick in firearms are often not being prosecuted for that conduct, rather they are being prosecuted for other related conduct.

Violations Reported in ATF Trafficking Investigations

Illegal buying and selling. ATF investigations uncovered a wide variety of apparent trafficking violations. ATF agents reported an average of two firearms violations per trafficking investigation and 36 percent of trafficking investigations were reported to have three or more firearms violations. Most were "selling" violations such as dealing without a license, or "buying" violations, such as straw purchasing, but violations involving firearms theft were also important.

Felons as traffickers. Felons play a significant role in firearms trafficking. About 23 percent of the investigations included violations involving convicted felons buying, selling, or possessing firearms.

Obliteration of serial numbers. About six percent of the investigations had violations involving serial number obliteration. Serial number obliteration is a clear indicator of firearms

trafficking, since the intentional obliteration of a serial number is intended to make it difficult for law enforcement officials to trace the firearm through a licensed seller to the first retail buyer. Other ATF analysis confirms that the obliteration of serial numbers is a significant problem; for eight cities with complete crime gun trace information, an average of 11.4 percent of traced handguns had obliterated serial numbers.[33]

Violations by FFLs. A review of the violations associated with the 133 investigations of licensed dealers revealed a variety of violations, including failure to keep required records (49 percent), transfers to prohibited persons (26 percent), making false entries in record books (19 percent), conducting illegal out-of-state transfers (16 percent), and obliterating serial numbers (8 percent).

---

[33] *Youth Crime Gun Interdiction Initiative Trace Analysis Reports, 27 Communities*, Department of the Treasury, Bureau of Alcohol, Tobacco and Firearms, February 1999. pg 13.

Table 14. Violations in ATF Trafficking Investigations*

Violation	Number of investigations with violation	Percent
Straw purchasing	638	41.7
Dealing without a license	486	31.8
Purchase/ possession/ sales by felon	350	22.9
Possession/ receiving/ trafficking stolen firearms	270	17.6
Non-licensee engaging in interstate firearms trafficking	244	15.9
Possession/ manufacture/ trafficking of NFA weapons	234	15.3
Burglary/ theft/ robbery of FFL/ common carrier/ manufacturer	205	13.5
Providing false information to acquire firearms	199	13.0
Trafficking - unspecified	135	8.8
International firearms trafficking	105	6.9
Obliterating or altering serial numbers	96	6.3
Sales to prohibited persons, including felons	89	5.8
Purchase/ possession/ sales by prohibited persons**	79	5.2
FFL failure to keep required records	66	4.3
Use of firearm during violent crime or drug trafficking	42	2.7
FFL transfer to prohibited persons	39	2.5
FFL making false entries in records	26	1.7
FFL conducts illegal out-of-state transfer	26	1.7
Trading firearms for drugs	26	1.7
Providing firearm to be used in violent/ drug crime	20	1.3
Trafficking/ possession by illegal alien	15	1.0
Providing firearms to juveniles	13	0.8
Non-licensee trafficking to felons	12	0.8
Transfer/ receive firearms through interstate commerce w/intent to commit felony	12	0.8
Armed career criminal	11	0.7
Failure to notify common carrier of firearms shipment	11	0.7
Possession/ manufacture/ transfer of illegal assault weapon	8	0.5
Other FFL violations (selling away from premises, failure to report multiple sales, failure to conduct Brady check, using false info to acquire FFL)	7	0.5
Possession/ manufacture of unregistered firearms	7	0.5
Unspecified violation(s)	8	0.5

Number of investigations included = 1,530

Note: Sum may exceed 100 percent since investigations may be included in more than one category.

 * Violations were observed but not necessarily presented as charges for prosecution.

** As distinct from felons; includes juveniles, illegal aliens, and other prohibited persons.

Violations in ATF Trafficking Investigations Involving FFLs

There were 133 ATF investigations (8.7 percent of 1,530) that involved the illegal diversion of firearms by a federally licensed firearms dealer (FFL). A variety of FFL violations were revealed in these investigations. Nearly half (48.9 percent, 65 of 133) of the licensed dealer investigations involved the failure to keep required records by an FFL. Slightly more than a quarter of the ATF investigations (26.3 percent, 35 of 133) involved FFLs transferring firearms to prohibited persons such as felons and juveniles. FFLs made false entries in their acquisition and disposition record book in almost a fifth of the ATF investigations (18.8 percent, 25 of 133) and conducted illegal out-of-state transfers of firearms in about 16 percent of the ATF investigations (15.8 percent, 21 of 133). ATF investigations involving corrupt FFLs were associated with many other types of firearms offenses including NFA violations (24.8 percent, 33 of 133), straw purchasing (24.1 percent, 32 of 133), and dealing without a license (18.8 percent, 25 of 133). Although corrupt FFLs are relatively rare, they are associated with the diversion of large volumes of firearms. While some violations committed by FFLs are felony offenses, two common FFL violations — failure to keep required records and making false entries in records — are misdemeanor offenses.

Table 15: Violations in ATF Trafficking Investigations Involving FFLs*

Violation	Number	Percent
FFL failure to keep required records	65	48.9
FFL transfer to prohibited persons	35	26.3
Possession/ manufacture/ trafficking of NFA weapons	33	24.8
Straw purchasing	32	24.1
Dealing without a license	25	18.8
FFL making false entries in records	25	18.8
FFL conducts illegal out-of-state transfer	21	15.8
Purchase/ possession / sales by felon	16	12.0
Trafficking - unspecified	14	10.5
International firearms trafficking	13	9.8
Sales to prohibited persons	13	8.8
Providing false information to acquire firearms	11	8.3
Obliterating or altering serial numbers	10	7.5
Possession/ receiving/ trafficking stolen firearms	9	6.8
Other FFL violations (selling away from premises, failure to report multiple sales, failure to conduct Brady check, using false info to acquire FFL)	7	5.3
Non-licensee engaging in interstate firearms trafficking	6	4.6
Trading guns for drugs	4	3.0
Receipt/ possession/ sales by prohibited persons	3	2.3
Use of firearm during violent crime or drug trafficking	3	2.3
Burglary/ theft/ robbery of FFL/ common carrier/ manufacturer	2	1.6
Providing firearm to be used in violent/ drug crime	2	1.5
Possession/ manufacture/ transfer of illegal assault weapon	2	1.5
Trafficking/ possession by illegal alien providing firearms to juveniles	1	0.8
Transfer/ receive firearms through interstate commerce w/intent to commit felony	1	0.8
Failure to notify common carrier of firearms shipment	1	0.8
Possession/ manufacture of unregistered firearms	1	0.8
Unspecified violation(s)	2	1.5

Number of investigations involving FFLs = 133

Note:　Sum may exceed 100 percent since investigations may be included in more than one category.

* Violations were observed but not necessarily presented as charges for prosecution.

National Firearms Act Violations

The National Firearms Act (NFA) regulates the manufacture, importation, transfer, and possession of specific types of firearms and other weapons that are considered particularly dangerous. Of the trafficking investigations, 15 percent involved NFA violations. Of these, 45 percent involved machine guns, 35 percent involved sawed-off shotguns and/or short-barrel firearms, 23 percent involved silencers, and about 17 percent involved firearms converted to fully automatic.

Table 16. Weapons Associated with NFA Violations in ATF Trafficking Investigations

NFA weapon	Number of investigations with at least one	Percent
Machine guns	105	44.9
Sawed-off shotguns/ short-barrel rifles & shotguns	82	35.0
Silencers	54	23.1
Converted guns	40	17.1
Conversion kits/ parts	21	9.0
Other explosives (e.g. blasting caps)	18	7.7
Grenades/ Grenade launchers	12	5.1

Number of investigations involving NFA violations = 234 (15.3 percent of 1530)

Note: Sum may exceed 100 percent since investigations may be included in more than one category. However, "converted guns" have not been included in the "machine gun" count.

Recommendations for Prosecution

Of the investigations initiated from July 1996 through December 1998, ATF agents recommended almost three-fourths of the firearms investigations for prosecution (1,090 of 1,530). Nearly 90 percent (962) of these 1,090 investigations were recommended to a U.S. Attorney's Office for prosecution, while slightly less than ten percent (101) were recommended to a State or local prosecutor for prosecution. Almost three percent (27) had some defendants in the case recommended to the U.S. Attorney's Office for prosecution and other defendants recommended to a State or local prosecutor for prosecution.

A variety of reasons were given by the ATF agents to explain why investigations (440) were not recommended for prosecution at the time of the supplementary survey. In less than 30 percent of the investigations not recommended for prosecution (121), the ATF agent stated that there was no potential for charges to be made against the defendant(s) in the case. Slightly less than 18 percent of the investigations (78) were not recommended for prosecution by ATF agents because there was insufficient evidence for prosecution at the time of the supplemental survey. Nearly 18 percent of the investigations not recommended for prosecution (77) were described as active ongoing investigations.

Table 17. ATF Trafficking Investigations Recommended for Prosecution

	Number of investigations	Percent
Investigations, total	1,530	100.0
Not recommended for prosecution	440	28.7
Recommended for prosecution	1,090	71.3
Recommended		100.0
to the U.S. Attorney's Office	962	88.2
to State/ local prosecutor	101	9.3
to both	27	2.5
Reason for not recommending	Number of investigations	Percent of 440
Total	440	100.0
No potential for charges	121	27.5
Insufficient evidence/ unable to substantiate violation	78	17.8
Investigation is still ongoing	77	17.5
Case was closed by ATF Field Division	34	7.7
Defendant(s) charged and prosecuted for other crimes before firearms trafficking investigation was developed	27	6.1
Investigation was transferred to another jurisdiction	16	3.6
Charges are forthcoming	9	2.0
Suspects were given a warning for violations without prosecution	8	1.8
Investigation was consolidated with another investigation	6	1.4
Confidential informant broke/ lost contact with defendant(s)	5	1.1
Investigation didn't progress because confidential informant was unreliable	4	0.9
No criminal intent by defendant(s) was found	4	0.9
Other reasons	23	5.2
Unspecified reasons	28	6.3

Status of Investigations Recommended for Prosecution

Ninety percent of the investigations (975 of 1,090) recommended for prosecution were accepted for prosecution; only 10 percent (115) were declined. Table 18 presents the status of the cases accepted for prosecution at the time of the supplementary survey. More than 57 percent of the investigations were fully adjudicated, while an additional five percent had some defendants fully adjudicated and other defendants not fully adjudicated. Thirty-six percent of the cases were not yet fully adjudicated and only two percent were on appeal.

Table 18. Status of Trafficking Investigations Accepted for Prosecution

Status	Number	Percent
Total	975	100.0
Fully adjudicated	558	57.2
Not fully adjudicated	351	36.0
Some defendants adjudicated, some not	46	4.7
Case is on appeal	20	2.1

Federal Charges Against Defendants Recommended for Prosecution in Trafficking Investigations

There were 1,787 defendants in the 1,090 firearms trafficking investigations recommended for prosecution by ATF agents. Almost 90 percent of these defendants faced Federal charges (1,593).

Trafficking-related charges. Some of the defendants were charged with offenses related on their face to firearms trafficking: 28 percent were charged with making false statements to acquire firearms, and about 18 percent were charged with dealing in firearms without a license. Nearly 28 percent of the defendants were charged with conspiracy to violate Federal laws and another eight percent were charged with aiding and abetting violations.

Other charges. Other defendants were charged with offenses not necessarily related on their face to firearms trafficking or firearms crime generally. Fully one quarter of the defendants were charged with being a convicted felon in possession of a firearm; another six percent were charged with other prohibited person violations. Over 27 percent were charged with conspiracy and over 12 percent with narcotics violations.

Table 19. ATF Description of Charges Against Defendants in Trafficking Investigations Recommended for Federal Prosecution

Charge category	Number of defendants	Percent
Conspiracy to violate Federal law	441	27.7
Making false statements to acquire firearms	439	27.6
Convicted felon in possession of firearm (includes Armed Career Criminal)	408	25.6
Dealing in firearms without a license	290	18.2
Theft of firearms/ Possession of stolen firearms/ Sale of stolen firearms	238	15.9
Illegal possession/ Transfer / Manufacture of NFA weapons	216	13.5
Narcotics violations	192	12.1
Aiding and Abetting in the violation of Federal law	122	7.7
User of a firearm during a drug crime	100	6.3
Possession of firearm by prohibited person (juvenile, drug user, illegal alien, restraining order, under indictment)	95	6.0
Interstate firearms trafficking	92	5.8
Sale of firearms to out-of-state resident	69	4.3
Possession/ shipping/ transporting a firearm with obliterated serial numbers	63	4.0
Failure to maintain ledger/ falsifying paperwork/ sale of firearms in violation of Brady law/ sale of firearms without noting purchaser's name, age, etc.	50	3.1
Selling firearms to convicted felon	35	2.2
Transfer of firearm to prohibited person (juvenile, illegal alien, restraining order, under indictment)	33	2.0
International firearms trafficking	28	1.8
Transfer of a firearm to be used in a violent crime	19	1.2
Other/ unspecified Federal violation	101	6.3

Number of defendants in the investigations recommended for prosecution = 1,593
Note: Since defendants can be charged with more than one type of crime, these categories are not mutually exclusive.

Geographic Distribution of Federal Prosecutions

Trafficking defendants were recommended for prosecution to U.S. Attorney's Offices in Federal courts covering 48 States, the District of Columbia, and Puerto Rico.

Table 20. States in which ATF Recommended Federal Prosecution of Defendants in Trafficking Investigations

State	Number	Percent	State	Number	Percent
Alabama	7	0.4	North Carolina	91	6.5
Alaska	4	0.3	North Dakota	16	1.0
Arkansas	5	0.3	Nebraska	8	0.5
Arizona	21	1.3	New Hampshire	11	0.7
California	54	3.4	New Jersey	17	1.1
Colorado	19	1.2	New Mexico	12	0.9
Connecticut	8	0.5	Nevada	10	0.6
DC	16	1.1	New York	134	9.4
Delaware	17	1.1	Ohio	37	2.3
Florida	97	6.1	Oklahoma	24	1.5
Georgia	98	6.2	Oregon	21	1.3
Hawaii	2	0.1	Pennsylvania	143	9.0
Iowa	7	0.4	Rhode Island	13	0.8
Illinois	24	1.5	South Dakota	9	0.6
Indiana	53	3.3	Tennessee	23	1.4
Kansas	6	0.4	Texas	124	8.9
Kentucky	23	1.4	Utah	2	0.1
Louisiana	23	1.4	Virginia	101	6.3
Massachusetts	43	2.7	Vermont	12	0.8
Maryland	24	1.5	Washington	29	1.8
Maine	17	1.1	Wisconsin	33	2.1
Michigan	25	1.6	West Virginia	27	1.7
Minnesota	4	0.3	Wyoming	2	0.1
Missouri	26	1.6			
Mississippi	13	0.8	Unspecified State	5	0.3
Montana	1	0.1			
			Puerto Rico	3	0.2

Number of defendants = 1,593

Disposition of Defendants Recommended for Prosecution

Adjudication. Nearly 61 percent of the defendants recommended for prosecution were fully adjudicated at the time of the supplemental survey (1,083 of 1,787). Only 3 percent of the 1,083 fully adjudicated defendants were found not guilty or were dismissed. The ATF agents knew the sentencing outcome for all but slightly more than seven percent of the sentenced defendants at the time of the survey (77 of 1,048).

Incarceration and fine. Nearly 78 percent of these defendants (812) received a term of incarceration. Slightly more than 7 percent of the incarcerated defendants also received a fine (60 of 812). The fines ranged from $50 to $40,000.

Term of incarceration. Of the 812 incarcerated defendants, 9 percent received a term that was more than 10 years with some receiving life imprisonment; 26 percent received a term that was more than four years but no more than 10

years; 26 percent received a term that was more than two years but no more than four years; 22 percent received a term that was more than one year but no more than two years, and 17 percent received a term of one year or less. The average sentence imposed on those incarcerated was about nine years of imprisonment, resulting in a total of 7,420 years in prison.

Probation. About 15 percent of the defendants who were sentenced received a term of probation (155 of 1,048). Of the 155 defendants who received probation, 23 percent received a term of more than four years but less than eight years, almost 33 percent received a term of more than two years but no more than four years; nearly 30 percent received a term that was more than one year but no more than two years; and nearly 13 percent received a term of one year or less.

Table 21. Disposition of Defendants in Trafficking Investigations Accepted for Prosecution

Status	Number of defendants	Percent
Accepted for prosecution	1,787	100.0
Not fully adjudicated	704	39.4
Fully adjudicated	1,083	60.6
Of those fully adjudicated	1,083	100.0
Guilty	1,048	96.7
Not guilty or dismissed	35	3.2
Disposition, total	1,048	100.0
Incarceration	812	77.5
Probation	155	14.8
Home confinement (w/o probation period)	4	0.4
Unknown sentence	77	7.3
Sentence length		
Incarcerated, total	812	100.0
12 months or less	138	17.0
13 – 24 months	176	21.7
25 – 48 months	207	25.5
49 – 120 months	210	25.9
121 months – Life	75	9.2
Unknown length	6	0.8
Fine in addition to incarceration	60	7.4
Probation, total	155	100.0
12 months or less	20	12.9
13 – 24 months	46	29.7
25 – 48 months	51	32.9
49 – 96 months	36	23.2
Unknown length	2	1.3
Fine in addition to probation	16	10.3
Home confinement in addition to probation	19	12.3

4. EXAMPLES OF GUN TRAFFICKERS AND THEIR SENTENCES

Residential FFL Trafficker

From August 1993 to March 1996, an FFL in Kansas City, Missouri, illegally sold 1,357 firearms, many from his van, over 200 of which were recovered at crime scenes in Kansas City. The firearms were primarily Lorcin and Bryco handguns. The FFL was charged with making sales in violation of State law; recordkeeping violations, and sale to a convicted felon. He entered a guilty plea on May 16, 1996, in the Western District of Missouri (Kansas City) and was sentenced on September 26, 1996, to 71 months' incarceration. This sentence was an upward departure; the sentencing guidelines called for a range of 37 to 46 months.

Gun Store Trafficker

Through Project LEAD, an investigation was initiated in July 1997, when two firearms, recovered in the Bronx, New York, were traced to an FFL in Suffolk County, New York. The investigation revealed that the FFL received a total of 207 firearms from a wholesaler that were then sold as part of an intrastate firearms trafficking operation. Thus far, five firearms have been recovered from crime scenes: four in New York City and one in Maryland. According to the FFL, the guns' serial numbers were removed by a co-conspirator in the Bronx, which accounted for the lack of additional known recovered firearms. In July, 1999, the FFL pled guilty to Criminal Sale of a Firearm, 2nd degree, a class "C" felony, in Suffolk County Court. On that same date, he was sentenced to 42 to 84 months in State prison.

FFL and Interstate Trafficker

This investigation began in March 1996 when a firearm recovered from a Washington, DC youth, charged with illegal possession of a firearm, was traced by the Washington, DC Metropolitan Police Department, after the ATF National Laboratory successfully raised the serial number. The paper trail led to a gun dealer in Missouri and later to a Nashville, Tennessee, gun trafficker who sold 200-300 guns on the streets of the Nation's capital. To date, 138 semiautomatic firearms originally sold by the Missouri Federal firearms licensee have been recovered in crimes in the Washington, DC. area ranging from nine murders to kidnapping, robbery, attempted murder, armed assault, drugs, and burglary. One of those firearms was used by a gang member to fatally wound a patient being transported to the hospital for treatment of wounds inflicted minutes earlier by the same gunman. On June 2, 1997, the Nashville gun trafficker pled guilty to Federal firearms trafficking charges. On August 22, 1997, he was sentenced to 60 months' imprisonment and 3 years' supervised release by the United States District Court in Nashville, Tennessee. During sentencing, the Federal judge referred to the defendant as a "dealer in death."

Straw Purchasers/Interstate Traffickers

This investigation was initiated in January 1998 by the ATF/ New York Police Department firearms trafficking task force. Between September 1997 and January 1998, two traffickers transported approximately 90 firearms from Georgia to New York. On February 5, 1998, members of the task force, accompanied by special agents from the Savannah, Georgia, field office, conducted surveillance on the subjects as they transported 11 firearms from Augusta, Georgia, to New York City. The two suspects were charged with illegally transporting firearms interstate and unlicensed dealing in firearms and conspiracy. One was sentenced on December 18, 1998, to 18 months' incarceration, and the other was sentenced on July 24, 1998 to 6 months' home detention and 5 years' probation.

Straw Purchaser/Intrastate Trafficker

This investigation was initiated in June of 1997 by ATF in Atlanta, Georgia. During a 9-month period, a straw purchaser had purchased more than 62 firearms in the Atlanta metropolitan area. These firearms were purchased for a defendant, who in turn sold the firearms on the streets of Atlanta. The straw purchaser was convicted and sentenced to 24 months to be served in boot camp. She testified in court that she had provided more than 200 firearms to the other trafficker. On June 5, 1998, the other trafficker pled guilty to violations of unlicensed dealing in firearms and possession of a firearm by a prohibited person, and was sentenced to 105 months in Federal prison.

Traffickers in Firearms Stolen From Gun Store Dealer

On November 17, 1996, more than 100 firearms were stolen from a pawn shop in Nashville, Tennessee. On November 18, 1996, two suspects were arrested when an Ohio State Police Highway Patrol Officer stopped their vehicle and recovered more than 73 of the firearms, as well as several other weapons. One suspect later pled guilty to possession of the stolen firearms.

Several of the remaining firearms that were trafficked by the suspects were recovered in separate incidents: In March 1997, an ATF investigation into the murder of a nightclub owner in Detroit, Michigan, resulted in the recovery of a Llama semiautomatic pistol that had been obtained by the victim. Subsequently, an associate of the murder victim surrendered another firearm, a Sturm Ruger 9mm semiautomatic pistol. Another of the pawn shop firearms (a Glock Model 26 9mm semiautomatic pistol) was recovered at the scene of a quadruple homicide in Nashville, Tennessee, on July 26, 1997.

On September 4, 1997, one of the two possessors of the stolen firearms received a pretrial diversion, later dismissed, by pleading to pos-

session of stolen firearms. The other possessor was sentenced to 57 months in prison and 3 years' supervised probation for violation of possession of stolen firearms and previously convicted felon in possession.

Traffickers in Firearms Stolen From Residences

The defendants were a group of young men who committed first commercial, then residential burglaries. The defendents stole entire gun safes out of homes and then cut them open inside a mini-storage unit they had rented. The suspects became increasingly bolder and more violent, ultimately committing a home invasion in which the female victim was tied up and her house robbed. Several stolen firearms were recovered during the ATF investigation, which involved approximately 200 firearms. Agents executed numerous search warrants which resulted in the recovery of several stolen firearms. Because of the fact that serial numbers were not available on most of the firearms reported as stolen from the residential burglaries, it is not known whether they have been recovered or not. The investigation revealed that the defendants traded these stolen firearms for narcotics or sold them for cash to unknown individuals. All defendants prosecuted were convicted or pled guilty and received a range of prison sentences from 33 to 300 months; two defendants were required to pay restitution of over $1 million.

Traffickers in Firearms Stolen From Common Carrier

On August 19, 1996, ATF special agents in Wilmington, Delaware, arrested a defendant for receipt of a large quantity of firearms stolen by another defendant, who had been arrested for the theft of 390 Llama firearms from an interstate shipment. Interviews with the thief yielded information regarding the identity and role of the recipient of the guns, who was subsequently arrested. (A third principal was arrested on August 14 for storing two cartons of the firearms.) On January 8, 1997, the recipient pled guilty to violation of possession of stolen firearms. He was sentenced to 33 months' incarceration on April 16, 1997. The thief was sentenced to 15 months' incarceration and 5 years' probation on January 20, 1998.

Straw Purchasers and International Traffickers

In 1996, ATF initiated an investigation following a seizure by the Colombian National Police of 17 MAK-90 firearms from a group of guerillas. The firearms were traced to several individuals in the Miami area, who had purchased 103 firearms from different licensed dealers and trafficked them to Colombia. Two suspects charged with making false statements were both sentenced to one year in prison, followed by deportation.

Licensed and Unlicensed Dealers Trafficking at Gun Shows

In 1996, ATF initiated an investigation of an unlicensed firearms dealer and convicted felon who was using straw purchasers to acquire large numbers of firearms for subsequent sale

to nonresidents. The primary suspect offered to sell agents an unlimited number of firearms, including fully automatic weapons and silencers.

Agents determined that the primary suspect had illegally trafficked more than 1,000 firearms during the past several years. Some of the firearms involved in this investigation were purchased at gun shows. One of the firearms was recovered from the scene of a shootout in Guadalajara in which two Mexican military officials were killed by drug traffickers. In 1997, another firearm was recovered during the search of the apartment of a former Mexican drug czar following his arrest for drug-related activities. Also under investigation were two federally licensed dealers suspected of selling stolen firearms, allowing individuals to straw purchase guns from them, and transferring firearms "off-the-book." Many of these transactions involved Mexican nationals.

In March 1998, the primary suspect was convicted of possession of an unregistered NFA firearm and sentenced to 78 months in prison, and an additional suspect, who was involved with straw purchasing firearms, was sentenced to 3 years' probation. In February and April 1998, the two licensed dealers were sentenced to 4 and 3 years' probation, respectively.

5. ENFORCEMENT LESSONS LEARNED FROM FIREARMS TRAFFICKING INVESTIGATIONS

ATF's review of trafficking investigations permits ATF to identify some of the important channels through which criminals are obtaining firearms, and provide insights into how they work. It also demonstrates that law enforcement can be successful in uncovering the illegal sources of firearms in a community and in holding gun traffickers accountable for their crimes. There are, however, challenges in enforcing the Federal firearms laws against gun traffickers.

5-1. Channels of Illegal Supply

ATF's review of recent trafficking investigations clearly demonstrates that many firearms are diverted from legal commerce through a variety of illegal channels that law enforcement agencies can target effectively to reduce criminal and juvenile access to firearms.

Trafficking by corrupt FFLs. Although FFL traffickers were involved in the smallest proportion of ATF trafficking investigations, under 10 percent, cases involving FFL traffickers were associated with the largest total number of illegally diverted firearms, over 40,000, as compared to the other trafficking channels.[34] All investigations involving FFL traffickers were associated with by far the highest average number of guns per investigation, over 350. But this number rose to over 560 guns in investigations of FFLs acting alone, and to over 575 guns when FFLs conspired with unlicensed sellers. Trafficking cases involved retail firearms dealers, pawnbrokers, and residential FFLs.

Gun shows and flea markets. Gun shows and flea markets are a major venue for illegal trafficking. About 14 percent of the investigations involved gun shows and flea markets. These investigations involved an average of 130 guns, the second highest number of trafficked guns per investigation, and were associated with approximately 26,000 illegally diverted firearms. Gun show investigations involved both FFL traffickers and unlicensed dealers.

Straw purchasers. Straw purchasing rings and small scale straw purchasers comprised nearly 50 percent of the trafficking investigations, by far the largest number of trafficking investigations, and although the average number of guns per investigation was under 40, they accounted for nearly 26,000 trafficked firearms, about the same number of firearms as gun show investigations. Straw purchasers may be friends, paid associates, relatives, intimates, or members of the same gang.

Trafficking by unlicensed sellers. Unlicensed sellers (not associated with gun shows and not straw purchasers) were the focus of about 20 percent of the investigations, involving over 22,000 trafficked firearms, and about 75 guns per investigation.

Trafficking in stolen firearms. Survey evidence indicates there are at least 500,000 firearms stolen annually from residences.[35] It is, therefore, not surprising that some of these firearms are circulating in the illegal market, as are guns stolen from FFLs and common carriers. Trafficking investigations involving firearms stolen from residences, FFLs, and common carriers combined made up about a quarter of the trafficking investigations, and were associated with over 11,000 trafficked firearms. Because of the small number of investigations involving

---

[34] Firearms may be trafficked along multiple channels; therefore, an investigation may be included in more than one of the categories described.

[35] Philip J. Cook and Jens Ludwig, *Guns In America*, Police Foundation, 1996.

---

thefts of firearms from common carriers, about 2 percent of the investigations, this trafficking channel yielded the smallest total number of firearms, although it averaged a substantial number of illegally diverted firearms per investigation, over 66.

Regional variations. Regional variations in the trafficking channels in ATF investigations suggest that the illegal market in guns may operate differently in different areas of the country. For instance, straw purchasing was involved in almost two thirds of ATF's trafficking investigations in its Northeast region, but closer to a quarter of its trafficking investigations in its Southwest and Western regions. Without additional information, it is not possible to know whether regional differences reflect differences in investigative practices, the illegal market, or both. ATF trace analysis also reveals differences in crime gun patterns among different cities.[36]

5-2. Impact of Enforcement

Convicted felons as traffickers. Although primarily aimed at the illegal supplier, about 45 percent of the trafficking investigations involved convicted felons in various roles. About 23 percent of the investigations included violations involving convicted felons buying, selling, or possessing firearms. Altogether, a quarter (669) of the traffickers identified in the investigations were convicted felons. A key finding of this report is that trafficking investigations lead to the apprehension of armed violent criminals, who are also important suppliers of the illegal market in guns.

Reducing criminal access to guns. These "criminals behind the criminal" were collectively responsible for the trafficking of over 84,000 firearms. Half of the trafficking investigations

involved guns known to be used in other types of crime, including felony possession, drug offenses, homicide, assault, and robbery. ATF estimates that trafficking investigations result in the actual seizure of about a quarter of the firearms ultimately shown to be trafficked by the targets. By arresting firearms traffickers, the number of firearms easily available to violent offenders can be reduced.

Enforcement successes. The Gun Control Act proscribes and penalizes illegal trafficking. ATF, supported by prosecutors, is enforcing the laws aggressively. ATF recommended almost three-fourths of the trafficking investigations for prosecution, nearly 90 percent to a U.S. Attorney, and the remainder to State and local prosecutors. Of these, 90 percent (over 1,700 defendants) were accepted for prosecution. Of the over 60 percent of the defendants who were fully adjudicated at the time of the survey, only 3 percent were found not guilty or dismissed. Over three-quarters of the defendants found guilty were sentenced to terms of incarceration, from sentences ranging from 12 months or less to life in prison. Altogether, as a result of these investigations, 828 defendants were sentenced to a cumulative 7,420 years in prison.

Necessity of regulatory enforcement. The major role played by FFLs as traffickers supplying the illegal market — from gun stores, pawnbrokers, gun shows, and residences, alone and in combination with straw purchasers and unlicensed sellers — confirms the need for a focused regulatory inspection and enforcement program based on trafficking and related indicators of potential illegal activity. Such a program has been initiated by ATF, and has been described in a report on regulatory enforcement published earlier this year by ATF.[37]

---

[36] *Crime Gun Trace Analysis Reports: The Illegal Youth Firearms Markets in 27 Communities,* Youth Crime Gun Interdiction Initiative, Department of the Treasury, Bureau of Alcohol, Tobacco and Firearms, February 1999.

[37] *Commerce in Firearms in the United States,* February 2000, Department of the Treasury, Bureau of Alcohol, Tobacco and Firearms.

5-3. Enforcement Issues

Systematic review of ATF's investigative experience led to some startling findings. While ATF agents and their State and local partners are successfully identifying firearms traffickers, U.S. Attorneys are accepting a high percentage of the cases, and juries are bringing in guilty verdicts, persons who traffick in firearms are often not being prosecuted for that conduct; instead, they are being prosecuted for other related conduct. Moreover, the penalties available for many corrupt FFLs who traffick hundreds of firearms, may be limited to misdemeanor recordkeeping violations. In addition, a trafficker who traffics 100 guns is subject to the same penalty as the straw purchaser who transfers 5 guns.

Obstacles to prosecution of straw purchasers and unlicensed dealers. Of the 638 investigations (41.7 percent of 1,530) where ATF agents reported that straw purchasing violations were occurring, a total of 427 were recommended for prosecution and not declined by the prosecutor (67 percent of 638; 51 were declined and 160 were still ongoing). The prosecutor charged at least one defendant with making false statements to acquire firearms in 190 of the 427 cases accepted for prosecution in which ATF agents reported straw purchasing was occurring.

Of the 486 investigations (31.8 percent of 1,530) where ATF agents reported that dealing without a license violations were occurring, a total of 296 of these investigations were recommended for prosecution and accepted by the prosecutor (61 percent of 486; 37 were declined and 153 were still ongoing). The prosecutor charged at least one defendant with dealing without a license in only 112 of the 296 cases accepted for prosecution in which ATF agents reported dealing without a license was occurring.

Thus, although ATF agents reported that straw purchasing and dealing without a license violations were occurring in the cases accepted for prosecution, the prosecutor was only able to charge at least one defendant with these violations in less than 45 percent of the straw purchasing cases and less than 38 percent of the dealing without a license cases. The fact that prosecutors accepted the cases shows that the conduct was viewed as constituting serious offenses. However, the relatively low percentage of charges facially tied to the trafficking conduct suggests that violations explicitly relating to firearms trafficking present challenges in court.

Challenges to enforcement against corrupt FFLs. The most common violation by FFLs associated with trafficking is recordkeeping violations. Failure to keep required records was found in almost half of the trafficking investigations involving FFLs, and the FFL making false entries in the records was found in just under a fifth of these investigations. These violations are primarily misdemeanors, despite being associated with investigations involving a high volume of trafficked firearms.

Limited penalties for large scale traffickers. The case review shows that while most trafficking investigations involve under 50 firearms, some investigations involve more than 50 firearms. Yet, the sentencing guidelines do not distinguish between a trafficker responsible for diverting 50 or 1,000 firearms. Thus, punishment can be disproportionately low given the gravity of the conduct and the threat to public safety. The more firearms illegally trafficked, the greater risk that some of those firearms will be used to commit serious crimes. The sentencing regime should be changed so that sentences are proportionate to impact. To address this concern, the Department of the Treasury has sought changes in the sentencing guidelines.[38]

---

[38] See letter from James E. Johnson, Under Secretary of the Treasury (Enforcement) to the Honorable Diane E. Murphy, Chair, U.S. Sentencing Commission, January 21, 2000; letter from James E. Johnson, Under Secretary of the Treasury (Enforcement) to Mr. John R. Steer, General Counsel, U.S. Sentencing Commission, November 17, 1999.

5-4. Investigative Partnerships and Resources

Role of State and local law enforcement. The case review reveals a rapidly developing capability to identify and apprehend gun traffickers. Most ATF trafficking investigations, 68 percent (1,037 of 1,530), involved the cooperation of State and local law enforcement agencies. This close working relationship seems especially appropriate because 70 percent of the trafficking investigations involved intrastate trafficking and about 10 percent involved trafficking in firearms stolen from residences. While slightly less than 10 percent of case referrals were to a State or local prosecutor, State and local trafficking enforcement can be expected to develop further with the growing use of firearms and ballistics trace information by State and local law enforcement agencies.

Firearms tracing. Nearly 20 percent of the trafficking investigations were initiated through trace analysis after the recovery of crime guns. Beyond the initiation of investigations, tracing was used as an investigative tool to gain information on recovered crime guns in 60 percent of the investigations.

Secondhand guns. The important role of tracing is reflected in the fact that a quarter of investigations involve only secondhand guns, while about 75 percent of the investigations involve at least some new guns. New guns are effectively traceable to the last retail purchaser, while secondhand guns are traceable only to the first purchaser, who may be several possessors removed from the source who supplied the gun to the criminal. ATF does not know what proportion of gun crimes are committed with new and secondhand firearms. It is clear, however, that secondhand firearms play a major role in firearms trafficking.

Continued growth of our ability to reduce the illegal market in firearms through enforcement activity clearly depends on expansion of our investment in comprehensive tracing of firearms and use of trace analysis, on investigative practices like the systematic debriefing of arrested violent offenders on the sources of their guns, and on the deployment of the National Integrated Ballistics Information Network (NIBIN), which can link spent cartridges and bullets to traceable firearms.

APPENDICES

APPENDIX A
Statutes Relating to Firearms Trafficking

Title 18, Section	Makes it illegal for	To	Mental Element(s)	Penalty, up to; Section
2	Any person	Aid, abet, counsel, command or solicit a criminal act	Knowingly, with the intent to assist in the commission of a crime	Same as for underlying crime
371	Any person	Agree with at least one other person to violate the law, with one person committing at least one overt act in furtherance of the agreement	True agreement with the intent to accomplish the objective(s) of the conspiracy	5 years if underlying crime is a felony; § 371
521	Criminal street gangs-5 or more persons whose primary purpose involves committing certain offenses.	Provides sentencing enhancement for gang members who have specified convictions within the previous 5 years.	Activities must affect interstate or foreign commerce.	Sentencing enhancement of up to 10 years; § 521.
922(a)(1)(A)	Any person	Engage in the business of dealing in firearms without a license	Willfulness; perpetrator must have acted with a bad purpose, either to violate the law or in conscious disregard of the law, but does not have to have known of the licensing requirement.	5 years; § 924(a)(1)(D)
922(a)(3)	Nonlicensee	Transport or receive firearm obtained in another state into his state of residence	Willfulness; not required to know he was breaking a specific law, but must have had bad purpose or motive	5 years; § 924(a)(1)(D)
922(a)(5)	Nonlicensee	Deliver firearm to unlicensed person whose residence is in a state different from transferor's	Willfulness; plus knowing or having reasonable cause to believe transferee resided in another state	5 years; § 924(a)(1)(D)

Title 18, Section	Makes it illegal for	To	Mental Element(s)	Penalty, up to; Section
922(a)(6)	Any person	Make a materially false statement to a licensee, to obtain or attempt to obtain a firearm	A lie is a knowing false statement. The statement must be "intended or likely to deceive" the licensee.	10 years; § 924(a)(2)
922(b)(2)	Licensee	Deliver firearm to a person where the purchase or possession of the firearm by that person would violate state law or published ordinance.	Willfulness	5 years; § 924(a)(1)(D)
922(b)(3)	Licensee	Deliver a firearm to a person residing in a state other than the licensee's	Willfulness; Also, licensee must know or have reasonable cause to believe that the transferee resides in another state	5 years; § 924(a)(1)(D)
922(b)(5)	Licensee	dispose of a firearm without making entries in records required to be kept under § 923.	Willfulness	5 years; § 924(a)(1)(D)
922(d)	Any person	Dispose of a firearm to a prohibited person	Transferor must know or have reasonable cause to believe the transferee is prohibited	10 years; § 924(a)(2)
922(g)	Any prohibited person	Possess or receive a firearm or ammunition that has moved in interstate or foreign commerce	Person must know he is in a prohibited category but need not know he is prohibited. Person must know he is in possession of a firearm.	10 years; § 924(a)(2)
922(i)	Any person	Transport or ship a stolen firearm in interstate or foreign commerce	Person must know or have reasonable cause to believe the firearm was stolen.	10 years; § 924(a)(2)
922(j)	Any person	Receive, possess, conceal, store, barter, sell or dispose of a stolen firearm that has moved in interstate or foreign commerce	Person must know or have reasonable cause to believe the firearm was stolen.	10 years; 924(a)(2)

Title 18, Section	Makes it illegal for	To	Mental Element(s)	Penalty, up to; Section
922(k)	Any person	Transport, ship or receive in interstate or foreign commerce a firearm with the serial number obliterated or altered, or to possess such a firearm that has been so transported or shipped	Knowing	5 years; § 924(a)(1)(B)
922(l)	Any person	Except as provided by § 925(d), to import or bring firearms or ammunition into the United States, or to possess such firearms or ammunition, or to receive such firearms or ammunition	"Knowing" as to the importation; "Willful" as to possession or receipt	5 years; § 924(a)(1)(C) and (D)
922(m)	Licensee	Make a false entry in, fail to make an entry in or fail to properly maintain records the licensee is required to keep	Knowing	1 years; § 924(a)(3)(B)
922(u)	Any person	Steal an "inventory" firearm from an FFL which has traveled in interstate or foreign commerce.	Knowing	10 years; § 924(i)(1)
924(a)(1)(A)	Any person	Make a false statement or representation in an FFL's required records or in applying for a license or relief from disabilities	Knowing	5 years; § 924(a)(1)(A)
924(b)	Any person	Ship, transport, or receive a firearm in interstate or foreign commerce	With the intent to commit a felony with the firearm or knowing or having reasonable cause to believe that a felony will be committed with the firearm.	10 years; § 924(b)

Title 18, Section	Makes it illegal for	To	Mental Element(s)	Penalty, up to; Section
924(c)	Any person	Uses or carries a firearm during or in relation to a Federal crime of violence or Federal drug trafficking crime; OR possess a firearm in furtherance of a Federal crime of violence or Federal drug trafficking crime.	Knowing	Minimum mandatory 5 years; 7 years if the firearm is brandished; 10 years if the firearm is discharged; additional enhancements for short-barrel rifles, short-barrel shotguns, machineguns, destructive devices, fire arms equipped with a silencer. Enhancements for second or subsequent conviction. Death penalty if death of a person results; § 924(c), 924(j).
924(h)	Any person	Transfer a firearm knowing it will be used to commit a crime of violence or a drug trafficking crime	Knowing for both elements	10 years; § 924(h)
924(k)	Any person	Smuggle or knowingly brings a firearm into the United States, or attempts to do so	Intending to engage in or promote conduct that violates state or Federal drug laws or that constitutes a crime of violence as defined in § 924(c)(3)	10 years; § 924(j)
924(l)	Any person	Steal a firearm that has traveled in interstate or foreign commerce	Theft is an intentional taking.	10 years; § 924(k)
924(m)	Any person	Steal a firearm from an FFL (no interstate commerce element)	Theft is an intentional taking.	10 years; § 924(l)

Title 18, Section	Makes it illegal for	To	Mental Element(s)	Penalty, up to; Section
924(n)	Any person	Travel from any state or foreign country into a state and acquires or attempts to acquire a firearm in such state	With the intent to violate 18 U.S.C. § 922(a)(1)(A)	10years; § 924(n)
1001	Any person	Make or use a false material statement in a Government-related matter	A lie is a knowing false statement. The statement must have been made for a bad purpose	5 years; § 1001
1341	Any person	Use the mail to perpetrate or advance a scheme to defraud	Having a scheme to defraud; Using the mail with specific intent to perpetrate or advance the scheme	5 years; 30 years if the scheme involves a financial institution. § 1341
2117	Any person	Burglarize a place or conveyance containing interstate or foreign shipments	Specific intent to commit a larceny therein	10 years; § 2117

APPENDIX B

Methodology

The case review presented in this report originated in a Congressional request for information about ATF's enforcement activities. Specifically, the House and Senate Committees on Appropriations requested ATF to report on trafficking investigations by February 1999 in connection with funding for ATF's Youth Crime Gun Interdiction Initiative (YCGII), the component of ATF's firearms enforcement programs focused on illegal acquisition, possession, and use of guns by youth and juveniles.[39]

In response, ATF Headquarters requested all ATF Special Agents in Charge to provide information on all firearms trafficking investigations in their respective areas between July 1996 (the commencement date of YCGII) and December 1998 (the end of the last calendar year before February 1999). A survey was sent to each Field Division requesting information for each investigation.

The 23 ATF Field Divisions submitted a total of 1,530 reports on investigations, including ongoing investigations and perfected cases referred for prosecution. Information on 648 investigations involving youth and juveniles were reviewed and provided the basis for a report to Congress on the performance of YCGII in February 1999.[40] In this report, ATF and an outside researcher review all 1,530 investigations.[41]

This report also reviews the disposition of cases referred by ATF for prosecution. To develop disposition information, ATF, in December 1999, sent supplementary surveys for the 1,530 submitted investigations to the 23 ATF Field Divisions. All surveys were returned to ATF Headquarters by March 15, 2000, for analysis by ATF personnel and outside researchers. Case disposition information was reviewed by outside researchers working with the Bureau of Justice Statistics, which has statutory authority for collecting and maintaining Federal case disposition information.[42]

---

[39] The Statement of Managers accompanying the 1998 Conference Report stated that: "the conferees believe that the proposed increase in funding must be supported by evidence of a significant reduction in youth crime, gun trafficking, and gun availability. The conferees would like to see additional evidence linking the Youth Crime Gun Interdiction Initiative (YCGII) to a corresponding decrease in gun trafficking among youths and minors. Therefore, the conferees direct ATF to report no later than February 1, 1999 on the performance of YCGII." Conference Report to Accompany H.R. 4328, October 19, 1998.

[40] See *Youth Crime Gun Interdiction Initiative: Performance Report.* Report to the Senate and House Committees on Appropriations Pursuant to Conference Report 105-825, October 1998. Department of the Treasury, Bureau of Alcohol, Tobacco, and Firearms, 1999.

[41] Dr. Anthony A. Braga of the John F. Kennedy School of Government, Harvard University.

[42] Dr. Anthony A. Braga of the John F. Kennedy School of Government, Harvard University and Dr. Joel Garner of the Joint Centers for Justice Studies, Sheperdstown, West Virginia.

Use of Surveys

ATF has a developing capacity to gather information from its Field Divisions. The analysis in this report is based on the best information currently available. ATF has conducted two previous case reviews, on gun shows and firearms trafficking.[43] These found that the survey methodology used here provided an accurate picture of the working knowledge held by agents involved in the investigations. A random sample of survey responses involving gun show investigations were carefully reviewed and compared to information contained in the investigation files. The investigation files contained a variety of information on the investigation, including a summary of the case, a set of progress reports documenting ATF agent investigative activities, police reports, evidence inventory, interview transcripts, and court documents. The review revealed that the survey responses were accurate when compared to the paperwork documenting the specifics of the gun show investigation. Another sample was drawn from the gun show

investigation data to verify the numbers of firearms reported as trafficked in each investigation.[44] Based on this review, the estimates of trafficked firearms made by the ATF agents were found to have a reasonable basis. Estimates were based on audits of firearms seized by the agents, firearms purchased by agents during the investigation, reconstruction of dealer records, ATF National Tracing Center crime gun recovery information, admissions by defendants, and information from confidential informants.

In addition, with the assistance of the Department of Justice, Bureau of Justice Statistics' Federal Justice Statistics Program, consistency was established between case disposition and sentencing information in the ATF survey and similar case disposition and sentencing information in automated records maintained by the Executive Office for U.S. Attorneys and the Administrative Office of the U.S. Courts.

What These Data Represent

Because these analyses are based on a survey of ATF special agents reporting information about firearms trafficking investigations, they reflect what ATF encountered and investigated in trafficking investigations. They do not look at the sources of firearms in other ATF investigations, such as investigations of armed career criminals or armed narcotic traffickers. Most importantly, they do not necessarily reflect typical criminal diversions of firearms or the typical acquisition of firearms by youth, juveniles, and adults. Except where noted, the unit of analysis in the review is the investigation itself.

ATF agents and their State and local counterparts gather investigative information to build a case worthy of prosecution, rather than to gather research information. Information generated as part of a criminal investigation, therefore, does not necessarily capture all data about trafficking, trafficking patterns, and the use of trafficked guns in crime. For instance, agents may provide very detailed descriptions of firearms used as evidence in the case but no estimate, much less a detailed description, of all the firearms involved in the trafficking enterprise. Thus, in general, agents did not provide enough consistent and specific infor-

---

[43] See US Department of Treasury and US Department of Justice, 1999. *Gun Shows: Brady Checks and Crime Gun Traces.* Washington, DC: US Department of Treasury and US Department of Justice.

[44] See US Department of Treasury and US Department of Justice, 1999. *Gun Shows: Brady Checks and Crime Gun Traces.* Washington, DC: US Department of Treasury and US Department of Justice.

mation to determine the number of handguns, rifles, and shotguns trafficked in a particular investigation. There may also be little information in the case file on the degree to which trafficked firearms were subsequently used in crime. This is primarily because comprehensive tracing of crime guns does not exist nationwide and, until the practice was initiated in 17 cities through the Youth Crime Gun Interdiction Initiative in 1996, even major cities did not trace all recovered crime guns. Figures on new, secondhand, and stolen firearms only reflect the number of investigations in which the traffickers were known to deal in these kinds of weapons. Figures on stolen firearms are subject to the usual problems associated with determining whether a firearm has been stolen, due to the fact that most gun owners do not report stolen firearms to the police. Such limitations apply to much of the data collected in this research.

Even the investigative data reported in the survey has limitations, because the review analyzes information both from investigations referred for prosecution and adjudicated, and from investigations not yet referred for prosecution. One third of the investigations were reported by the agents as fully adjudicated (514 of 1,530). Not all violations described will necessarily be charged as crimes or result in convictions, and case agents at the time of the survey might not know the exact number of offenders in the investigation, the numbers and types of firearms involved, and the types of crimes associated with recovered firearms. Some information may have been underreported. For example, it is likely that the number of firearms involved in the investigations could increase, as could the number and types of violations, as more information is uncovered by agents working the investigations.

SURVEY FORMS

ATF Field Division:

ATF Investigation Number:

Investigation Title:

1. How was the investigation initiated?

1= confidential informant

2= referral from another state, local, or federal agency

3= FFL reported suspicious activity

4= developed from another investigation

5= review of multiple sales forms

6= Project LEAD or other local tracing project

7= gun recovered and traced to origins

8= ATF initiated investigation of suspicious activity (e.g. ATF Gun Show Task Force)

9= ATF Regulatory inspection of FFL records

10= anonymous tip

11= other Please specify:_____

2. What were the violations in the investigation?

Please use the space below to write all violations associated with the traffickers in the investigation (e.g., trafficking, straw purchasing, dealing by non-licensed individuals, "off paper" sales, sales to prohibited persons, obliterating firearms, etc.)

3. Were there any Title II violations in this investigation?

0= NO   1= YES

Please use the space below to describe the Title II violations (e.g., machine guns, converted guns, conversion kits, silencers, grenades, short barreled firearms, etc.).

4. Please describe the trafficking enterprise.

Circle all appropriate trafficking channels listed below. For example, if a FFL was trafficking firearms at a flea market, choices 1 and 7 would be selected.

1= Firearms trafficked by licensed dealer, including pawnbroker.

2= Firearms trafficked by straw purchaser or straw purchasing ring.

3= Trafficking in firearms stolen from FFL.

4= Trafficking in firearms stolen from common carrier.

5= Trafficking in firearms stolen from residence.

6= Trafficking in firearms by unregulated private sellers.

7= Trafficking in firearms at gun shows, flea markets, auctions, or want ads and gun magazines.

8= Other.   Please specify:_____

5. Were there NEW guns being trafficked?

0= NO   1= YES

6. Were there OLD guns being trafficked?

        0= NO   1= YES

7. Were there USED guns being trafficked?

        0= NO   1= YES

8. Were the firearms being trafficked INTERSTATE?

        0= NO   = YES

9. Were the firearms being trafficked INTRASTATE?

        0= NO   1= YES

10. Were the firearms being trafficked INTERNATIONALLY?

        0= NO   1= YES

11. Please describe the number and types of firearms involved in the investigation.
Please provide as much detail as possible; if the number of handguns and long guns are not known, only fill in the number of firearms.

Total number of firearms: _____

Total number of handguns: _____

Total number of long guns: _____

Total number of rifles: _____

Total number of shot guns: _____

Beyond the number of firearms known to be involved in the investigation, please estimate the number of firearms that may have been trafficked by the individual(s) under investigation.

Estimated firearms: _____

12. Did this investigation involve YOUTH (person ages 18-24)?

0= NO   1= YES

If yes, how were youth involved? Circle all that apply

1= straw purchaser(s)

2= trafficker(s)

3= possessor(s) of trafficked firearms

4= thief/robber(s)

5= other.   Please specify:_____

13. Did this investigation involve JUVENILES (person ages 17 and under)?

0= NO   1= YES

If yes, how were juveniles involved? Circle all that apply

1= straw purchaser(s)

2= trafficker(s)

3= possessor(s) of trafficked firearms

4= thief/robber(s)

5= other.   Please specify:_____

14. Were the trafficked firearms known to be recovered in subsequent crimes?

0= NO   1= YES

If yes, please enter the number for each crime below. If exact numbers are unknown, please simply put a check in the space provided.

Homicide: _____

Assault: _____

Robbery: _____

Sexual assault/ rape: _____

Property crime: _____

Felon in possession: _____

Juvenile in possession: _____

Illegal possession: _____

Drug offense: _____

Other (please specify):   _____

15. Please provide information on the defendant(s) in the investigation. Please use the space below to describe their role in the case (e.g. straw purchaser, FFL, trafficker, etc.), age, sex, race, and whether the person had a felony record.

16. If there were straw purchasers involved in this investigation, please provide any information you may have on the relationship between the straw purchaser and the actual trafficker. Please circle all that apply.

1= Friend

2= Intimate

3= Relative

4= Business relationship (paid with money or drugs to buy guns for the trafficker)

5= Straw purchaser is the trafficker.

6= Other

Please specify: _____

17. Has this case been adjudicated?

   0= NO   1= YES

18. Please use this page to summarize any other pertinent information on this investigation.

Bureau of Alcohol, Tobacco, and Firearms
Supplementary Survey of Recent Gun Trafficking Investigations

INSTRUCTIONS:

We would like to know some additional information regarding the disposition of the firearms trafficking investigation listed below that your Field Division submitted to ATF Headquarters. Please select the appropriate response or write your response to the following questions in the space provided. Thanks for your time and consideration in this important matter.

Field Division:

ATF Investigation Number:

ATF Investigation Title:

1)   Was this investigation recommended for prosecution?   _____ YES   _____ NO

2)   Who was the investigation recommended to for prosecution?

_____ US Attorney's Office

_____ State/ local prosecutor

If the investigation was not or has not been recommended, please briefly state why.

3)   If this investigation was recommended for prosecution, was the submission declined by the prosecutor?

_____ YES   _____ NO

4)   If the investigation was accepted for prosecution, please list the <u>defendants</u> (name, DOB) in the case and the <u>charges</u> under which they were prosecuted.

5)   Has the case been fully adjudicated?

_____ YES   _____ NO   _____ ON APPEAL

_____ Some defendants have been adjudicated, some not

For the adjudicated defendants, please list the sentence for each individual.

6)   Please name the prosecutor(s) of the case and identify which U.S. Attorney District or local court he/she works.

7)   Did this investigation involve the cooperation of a local or state police department?

_____ YES   _____ NO

8)   Was the tracing of firearms used in this investigation?
_____ YES   _____ NO

If yes, what was the role of tracing? (please select all that apply)

_____   Investigation was initiated through firearms tracing

_____   Recovered firearms were traced after the investigation was initiated

_____   Other, please describe: